D0889825

Masers and Lasers

An Historical Approach

Frontispiece. Dr T H Maiman with the first ruby laser.

Masers and Lasers

An Historical Approach

M Bertolotti

Institute of Physics
University of Rome

Adam Hilger Ltd, Bristol

British Library Cataloguing in Publication Data

Bertolotti, M
Masers and lasers.
I. Title
621.36′6 TA1675

ISBN 0-85274-536-2

Consultant Editor: Dr E R Pike RSRE, Malvern

Published by Adam Hilger Ltd, Techno House, Redcliffe Way, Bristol BS1 6NX.

The Adam Hilger book-publishing imprint is owned by The Institute of Physics.

Typeset by Unicus Graphics Ltd, Horsham, and printed in Great Britain by Pitman Press Ltd, Bath.

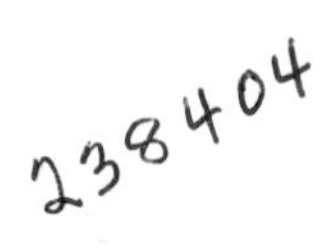

Contents

Foreword

Masers, and especially lasers, are by now familiar devices that are very widely used in many fields of science and technology. To some, lasers seem to offer a fulfilment of one of mankind's oldest dreams of technological power, an all-destroying energy ray. That may be the basis for the ancient, probably apochryphal, legend that Archimedes was able to set enemy ships on fire by using large mirrors to reflect and focus sunlight. The English novelist H G Wells, in his 1898 story *War of The Worlds* had Martians nearly conquering the Earth with a heat ray. In the 1920s, the Soviet novelist Alexei Tolstoi wrote *The Hyperboloid of Engineer Garin*, which device also was described as producing an intense light beam. Newspaper comic strips such as *Buck Rogers* in the 1930s made 'disintegrator guns' familiar ideas.

Yet as scientists learned more about how light is produced and absorbed, these novelists' dreams seemed all the more unlikely. Thermal emitters of light, which were all we had, seemed to absorb the light that they emitted, if we tried to make them thicker to get more intensity. Yet it turned out that this and many other apparent difficulties could be overcome. Indeed, when the first lasers were operated, I and other scientists close to the research were surprised at how easy it turned out to be. We had assumed that, since lasers had never been made, it must be very difficult. But once you knew how, it was not at all difficult. Mostly what had been lacking were ideas and concepts.

This book blazes a new trail, in retracing the history and expounding the theory and experiments as they were discovered. This is a complex task, as there is no earlier book of comparable scope to use as a starting point. Inevitably, there are many points in the discussion which I would state differently, and some which I would have to dispute. But Professor Bertolotti's long experience in this field, to which his research has contributed much, has enabled him to produce a sound outline of the way things developed.

This treatment could also serve very well as an introduction to the theory of masers and lasers, since these matters are thoroughly discussed in a sequence appropriate to the historical presentation.

Arthur L Schawlow
Department of Physics
Stanford University
26 October 1982

Professor A L Schawlow with an early ruby laser cooled by liquid nitrogen.

Preface

Nowadays masers and especially lasers are very popular devices. When masers were invented in 1951, they offered a completely new and revolutionary method for producing microwaves. The theoretical foundations necessary to understand the way they work and actually to build them, however, had already been well established in the 1930s. It took people 20 years to get rid of the old traditional schemes of producing electromagnetic waves and to find a completely revolutionary path of achieving them and, as often occurs in these cases, at that moment the time was ripe and the same idea occurred to many scientists almost simultaneously.

Lasers were the natural extension of this idea to light. Mostly popularised, invented even before they actually worked by science-fiction writers as the 'death ray' and other similar names, lasers were for some time ironically defined as a 'solution in search of a problem'.

Nowadays there is scarcely any physical laboratory which does not own at least one. More applications are discovered every day. In fact lasers have not been fully exploited yet, and probably their best applications are to come.

I think it both interesting and instructive to trace out the history of how these devices developed since the first basic principles at their origin were established.

And now let me say, as Agamemnon does in Troilus and Cressida by William Shakespeare (Act I, Scene III):

> Speak, Prince of Ithaca; and be't of less expect,
> That matter needless, of importless burden
> Divide thy lips, than we are confident,
> When rank Thersites opes his mastiff jaws,
> We shall hear music, wit, and oracle.

Before starting let me first acknowledge help from many people who provided me with documentation and discussion. Notably Professors N Bloembergen,

B Crosignani, P Di Porto, H Gamo, R J Glauber, S F Jacobs, B A Lengyel, V S Letokhov, S I Nishira, A M Prokhorov, M Sargent, A L Schawlow, C P Slichter, C H Townes and V Vavilov.

Special thanks are due to Dr Roy Pike who kindly turned my Anglo-Italian into English, to Miss A De Cresce who, with great patience and ability, typed the manuscript and to Mrs F Medici and C Sanipoli for preparing drawings and diagrams.

1

Introduction

This short book aims to trace out a history of the development of the fundamental ideas on which masers and lasers are based.

Although this development has been deeply impressed on the minds of all of the early researchers in this field, it may well escape the knowledge of young researchers approaching this area for the first time. Graduate students in physics and electronics can profit from such knowledge, and the level of exposition has been chosen accordingly. Whenever possible original authors have been allowed to speak for themselves through their papers, and short biographies of the leaders in the field have been included.

1.1 Principle of operation

Masers and lasers have in common their principle of operation, which is based on the use of stimulated emission of electromagnetic radiation in a medium of molecules or atoms with more particles in the upper (*excited*) state than in the lower state (that is, with an *inverted population*).

An electron bound to a molecule or an atom may change its energy state by jumping from one energy level to another with the emission or absorption of a photon, in which process it will have, respectively, lost or gained energy.

If the particle is initially in an upper state with respect to the fundamental ground state, it decays spontaneously to the ground state by the emission of a photon of energy

$$h\nu = \Delta E, \tag{1.1}$$

where ΔE is the energy difference between the two levels, ν the frequency of emission, and h Planck's constant.

This is the normal emission process (*spontaneous emission*) which takes place every time a material de-excites itself after having been suitably excited. Due to the random nature of spontaneous decay, photons are emitted by the various particles in an independent way and the resultant emission is incoherent.

The probability of spontaneous emission increases with the cube of frequency. It is therefore negligible in microwave transitions where, instead, thermal relaxation processes are predominant.

If electromagnetic radiation is present with a frequency such as to fulfil equation (1.1), two processes can be distinguished:

(a) the photon interacts with a particle which is in its lower energy level: in this case radiation is absorbed and the particle is forced to go to the upper level;

(b) the photon interacts with a particle which is already in an upper state. In this case the particle is forced to go to the lower energy state by emitting another photon of the same frequency as the incident photon. This process is called *stimulated emission*.

While with spontaneous emission both the direction and polarisation of photons are randomly distributed, in the case of stimulated emission they coincide with those of the incident photon.

In general, the particles of an ensemble in equilibrium are more numerous in the lower energy level, according to the Maxwell-Boltzmann distribution law. By making the system interact with radiation of frequency equal to the difference between two levels (the ground and the excited levels) processes of type (a) and (b) take place simultaneously, with a prevalence of the former because more particles are in the lower state. However, if, in some way, the distribution between levels is altered, so that more particles are present in the excited level than in the ground state (one usually says that an *inversion of population* has been obtained), then in the interaction process between the radiation and the particles a net excess of emitted photons will take place over the absorbed photons, i.e. an amplification process occurs.

The radiation emitted in this way will be monochromatic – because it is emitted in correspondence with a well defined transition – and coherent – because it is a forced emission in phase with the driving field.

This is the fundamental working principle common to masers and lasers.

1.2 The devices

Masers are devices emitting in the microwave region and, therefore, molecular roto-vibrational states are usually used. Lasers are devices emitting in the region from infrared to ultraviolet and they can, therefore, use either molecular or atomic levels.

Once suitable energy levels have been chosen, the next problem with which one is faced is how to obtain an appreciable amount of radiation and how then to couple that radiation efficiently to the particles. This is achieved by the use of a suitable resonant cavity, i.e. a microwave cavity for masers and an optical cavity for lasers. The effect of such a cavity is not only to increase the residence time of the photons in contact with the ensemble of excited atoms so as to increase the probability of de-excitation, but also to provide the feedback necessary to make the emitted wave grow coherently.

The logical steps in building up such devices are consequently:

(a) the understanding of the role played by stimulated emission and thermodynamical equilibrium conditions in an ensemble of particles interacting with an electromagnetic field;
(b) the creation of a suitably inverted population; and
(c) the appreciation of the role of a resonant cavity.

Once these three basic features were well integrated into one single concept, first masers and later lasers could be constructed.

Masers

In masers one may consider two fundamental methods for inverting a population:

(1) A preselection of excited molecules in a gas. This was the first method used by Townes, in which an excess of excited molecules from ammonia gas was introduced into the resonant cavity. A very small signal was therefore greatly amplified at the molecular transition. This kind of maser is very suitable as a low noise single frequency amplifier or as a frequency standard.

(2) The use of paramagnetic levels in a solid material. This was a later proposal made by Bloembergen. The three-level maser is the best example of this type: its working principle is shown with reference to figure 1.1. A suitable paramagnetic ion is put into a magnetic field so that a set of Zeeman levels is created. Let us consider three of these levels. With a suitable *pump* frequency, particles are excited from the lower, ground level 1 to the upper level 3. If the energy between levels is large compared with kT, level 2 is approximately empty and therefore an inversion of population is created between levels 3 and 2. If the ions are put into a resonant cavity, amplification of radiation at frequency ν_{32} occurs.

The distance between levels is a function of the strength of the external magnetic field. Therefore the frequency ν_{32} can, to some extent, be tuned. This maser is therefore much more flexible than masers of type (1).

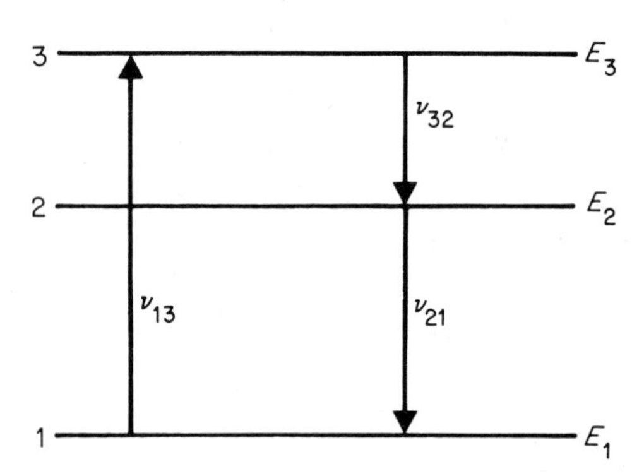

Figure 1.1 The three-level diagram proposed by Bloembergen for a solid-state paramagnetic maser.

Lasers

Lasers were developed later. Initial proposals were for gas media, but the first realisation was in a solid. This was followed by devices exhibiting laser operation in a myriad of different systems.

The most important obstacle to the progression from masers to lasers was the choice of a suitable resonant cavity at optical frequencies. The solution proved to be an open cavity in the form of a Fabry–Perot device, two parallel plane mirrors some distance apart from each other. Light travelling along the axis of the system is forced to reflect back and forth, the principal condition being that in the distance of separation between the two mirrors there is an integral number of wavelengths.

The principle of operation of a laser is the same as for masers. For example, in the ruby laser, which was the first to be successfully operated, the excited population is produced by illumination of Cr ions in corundum with a broad band of visible light from a flash lamp. The light in the green part is absorbed by the ions which are then excited into a band of higher levels. From this band they decay, without emission of light, simply by giving their excess energy to the lattice as heat, to an intermediate level (actually a doublet) which can therefore become more populated than the fundamental level.

From this intermediate level (or rather, from one of the two doublet levels) Cr ions finally decay via strong light emission at 6943 Å at room temperature to the fundamental level, and this is the radiation which is amplified in a suitable Fabry–Perot cavity.

1.3 Applications

In figure 1.2 the chronology of coherent wave generation is shown – an impressive visual display. The frequencies obtained are seen to follow an approximate exponential law up to the invention of lasers. From that point on, the trend has a marked change in slope which predicts the attainment of very short wavelengths to be rather delayed in time.

In the case of microwaves, the maser followed many other devices for their production. It has, however, unique properties, mainly high monochromaticity and an extremely low noise temperature.

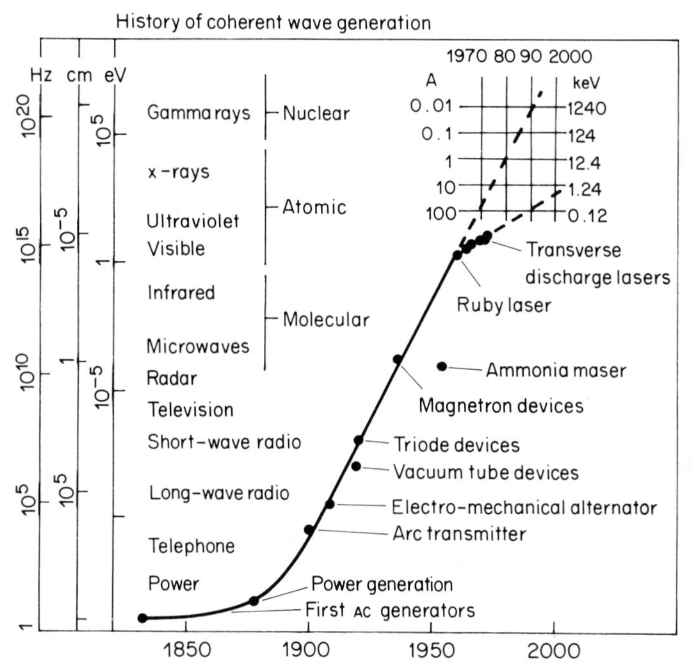

Figure 1.2 History of coherent wave generation (by courtesy of Dr D J Nagel, Naval Research Laboratories, Washington DC).

However, in the case of the infrared–visible–ultraviolet region the laser constituted a big revolution. With it, for the first time, man was able to have in the optical region a fully coherent source with enormous brightness and monochromaticity.

By just using these two properties a number of fields in optics received a large boost. Holography – which had already been invented by Gabor in 1948 – and non-linear optics – which was just the optical counterpart of well known phenomena in radiowaves – were two domains, for example, in which rapid growth was made practicable by the availability of the laser. Two other fields which, although well known and in operation well before the advent of the laser, received by its use a new and extraordinary boost were interferometry and all light-scattering techniques.

However, the revolution was not complete until it was appreciated that the properties of the light emitted by lasers were completely different from those of light emitted by an ordinary source, such as a lamp or an ordinary gas discharge.

The use of stimulated emission in an inverted population is a way of producing radiation in conditions far from equilibrium and therefore the statistical properties of the emitted light are completely different from those of any other light source known before. This point, which initiated a new activity in coherence theory, has importance, for example, in scattering and detecting light.

1.4 Development

In this book an effort is made to retrace the conceptual path which has brought the invention of first masers and then lasers.

The history of the development of these devices can be divided roughly into four periods. The first period starts with the introduction of the concept of stimulated emission and ends with its experimental demonstration. It extends from 1916 to 1953, and it is covered by Chapters 2 and 3.

The second period begins with the construction of the first maser in 1954 and ends in 1960 with intensive speculation concerning the extension of the art of the maser from the microwave to the optical region. This maser period is covered by Chapters 4 and 5.

The laser or third period opens with the achievement of the first operating laser. It is characterised by an explosive growth of research into and development of many different kinds of laser. This period extends from 1961 to roughly 1970.

The last or second-generation laser period is characterised by the development of many different kinds of laser which have the characteristics necessary to make them of use for industrial, medical and general technical research. It opens about 1970 and is still open today. These last two periods are covered in Chapter 6. The subject of applications is omitted, since this would have more than doubled the space required. Chapter 7 is instead devoted to the statistical properties of laser light and to the latest developments in coherence theory, in the light of the previous discussion.

2

Stimulated emission: Could the laser have been built more than 50 years ago?

2.1 Stimulated emission

In 1916 Einstein (1879-1955)[1] published a new, extremely simple and elegant proof of Planck's law of radiation and, at the same time, obtained important new results concerning the emission and absorption of light from atoms or molecules. In his paper, for the first time, the concept of *stimulated emission*, which is basic to the laser effect, is introduced. Our story can therefore begin by analysing the methods used and conclusions reached by Einstein. He skilfully combined 'classical laws' with the new concepts of quantum mechanics, which were at that time growing up. The line of reasoning he followed was more or less similar to the one adopted by Wien (1864–1928) in his derivation of the radiation law[2]; Einstein, however, adapted it to the new situation created by Bohr's spectral theory. At the beginning of his paper he writes 'Not long ago I discovered a derivation of Planck's formula which was closely related to Wien's original argument[2] and which was based on the fundamental assumption of quantum theory', and later on, 'I was led to these hypotheses by my endeavour to postulate for the molecules, in the simplest possible manner, a quantum-theoretical behaviour that would be the analogue of the behaviour of a Planck resonator in the classical theory'.

His reasoning is the following. Let us consider some well defined molecule, without taking into account its orientational and translational movements. According to the postulates of the quantum theory already developed at that time, it could have only a discrete set of states $Z_1, Z_2, \ldots, Z_n$, whose internal energies can be labelled with $\epsilon_1, \epsilon_2, \ldots, \epsilon_n$. If a large number of such molecules belong to a gas at temperature T, the relative frequency W_n of the state Z_n is given by the formula of Gibbs' canonical distribution modified for discrete states, i.e.[3]

$$W_n = g_n \exp(-\epsilon_n/kT),$$

where g_n is a number, independent of T and characteristic for the molecule and its nth quantum state, which can be called the *statistical weight* of this state.

Let us now take into account the radiative exchanges of energy (i.e. processes with emission or absorption of electromagnetic waves). They are treated by considering a Planck resonator in the classical scheme. Einstein identifies three different interaction mechanisms. The first one is a radiative emission process which we now call *spontaneous emission*. To describe this process he assumes that the probability of a single molecule in state Z_m going in a time dt, without being excited by external agents, to the state of lower energy Z_n, with the emission of radiant energy $\epsilon_m - \epsilon_n$, to be

$$A_m^n \, dt, \tag{2.1}$$

where A_m^n is a given constant.

The jumps from one energy level to another that in the Bohr theory give rise to the spontaneous emission of radiation are taken as being analogous to spontaneous radiactive disintegrations, and Einstein in writing equation (2.1) assumed that radiative transitions of free atoms are governed by a probability law similar to the one postulated in the elementary theory of radioactivity. τ_m^n, the reciprocal of A_m^n, is the spontaneous lifetime of the upper level with respect to the lower one.

He then considers the absorption of radiation and writes[4]:

> If a Planck resonator is located in a radiation field, the energy of the resonator is changed through the work done on the resonator by the electromagnetic field of the radiation; this work can be positive or negative, depending on the phases of the resonator and the oscillating field. We correspondingly introduce the following quantum-theoretical hypothesis. Under the influence of a radiation density ρ of frequency ν a molecule can make a transition from state Z_n to state Z_m by absorbing radiation energy $\epsilon_m - \epsilon_n$, according to the probability law
>
> $$dW = B_n^m \rho \, dt. \tag{B}$$
>
> We similarly assume that a transition $Z_m \rightarrow Z_n$, associated with a liberation of radiation energy $\epsilon_m - \epsilon_n$, is possible under the influence of the radiation field, and that it satisfies the probability law
>
> $$dW = B_m^n \rho \, dt, \tag{B'}$$
>
> B_n^m and B_m^n are constants. We shall give both processes the name 'changes of state due to irradiation'.

Process (B′) is the one we now call *stimulated emission* (induced emission) and is introduced here for the first time.

The term 'stimulated emission' does not appear in the above quotation and was introduced at a later time by J van Vleck (1899–1980) in 1924[5].

Once these fundamental hypotheses have been established, Planck's law is derived at once by assuming the energy exchange between radiation and molecules does not perturb the canonical distribution of states given before.

Therefore, averaged over unit time, as many elementary processes of type (B) as of emission (2.1) and (B′) must take place

$$g_n[\exp(-\epsilon_n/kT)]\,B_n^m\rho = g_m[\exp(-\epsilon_m/kT)]\,(B_m^n\rho + A_m^n). \qquad (2.2)$$

Let us now assume that, by increasing T, ρ also increases tending towards infinity. From equation (2.2) we obtain

$$g_n B_n^m = g_m B_m^n, \qquad (2.3)$$

and

$$\rho = (A_m^n/B_m^n)/\{\exp[-(\epsilon_m - \epsilon_n)/kT] - 1\}. \qquad (2.4)$$

This is Planck's radiation law. It gives asymptotically Rayleigh's law for large wavelengths and Wien's law for small wavelengths if we take

$$\epsilon_m - \epsilon_n = h\nu, \qquad (2.5)$$

and

$$A_m^n = (8\pi h\nu^3/c^3)\,B_m^n. \qquad (2.6)$$

In this and in all other derivations up to the time of Bose's paper on statistics in 1924, the proportionality factor between A and B was obtained by appeal at one point or another to classical electromagnetic theory. The quantity $8\pi\nu^2/c^3$ represents the number of normal modes of the radiation per unit volume and per unit frequency interval[6].

Equations (2.3) and (2.6) appeared for the first time in this work by Einstein and are fundamental to the theory of energy exchange between matter and radiation[7]. The probabilities that Einstein assumed for each of the elementary processes suffered by a molecule are today indicated as *transition probabilities* between states. Bohr's quantum theory did not give any indication of the laws governing such transitions and the concept of transition probability originated in Einstein's paper.

One of the principal problems of quantum mechanics at that time was to calculate these coefficients from data pertaining to atoms and molecules.

Equation (2.6) was experimentally verified through a comparison of the intensities of absorption and emission lines. The constants B were obtained from measurements of the intensity of multiplet components in spectra by L S Ornstein and H C Burger[8].

Another important result established in Einstein's work is connected with the exchange of momentum between atoms or molecules and the radiation. Einstein showed here that when a molecule (atom), making a transition from state Z_n to Z_m receives energy $\epsilon_m - \epsilon_n$, it also receives momentum $(\epsilon_m - \epsilon_n)/c$ in a defined direction. Moreover, when a molecule (atom) in the transition from Z_m to the lower energy state Z_n, emits radiant energy $\epsilon_m - \epsilon_n$, it gains momentum $(\epsilon_m - \epsilon_n)/c$ in the opposite direction. Therefore the emission and absorption processes are *direct processes*; emission or absorption of spherical waves is not likely to occur. He writes 'Outgoing radiation in the form of spherical waves does not exist'[9].

The subsequent Compton (1892–1962) experiment on the scattering of x-rays provided the first experimental confirmation of these predictions[10].

Einstein's theory of emission and absorption allowed W Bothe (1891–1957)[11] to give an instructive calculation of the number of radiation quanta $h\nu$ of a black body that are associated as 'photo-molecules' in pairs $2h\nu$, triplets $3h\nu$, etc. He considered a cavity filled with black-body radiation at a temperature T in which a large number of gas molecules were enclosed, each one being in one of the two states Z_1 and Z_2, where $Z_2 - Z_1 = h\nu$, their relative mean numbers being given by the canonical distribution law at temperature T. He assumed that when a single quantum $h\nu$ of the radiation causes a stimulated emission the emitted quantum moves with the same velocity and in the same direction as the stimulating quantum, so that they become a pair of quanta $2h\nu$. If the exciting quantum itself already belongs to a pair, then a triplet $3h\nu$ is produced, and so on.

The absorption of one quantum by a pair of quanta leaves a single quantum, and spontaneous emission produces a single quantum. By writing the conditions for the mean number of single quanta, of quantum pairs, etc to be constant in time and by using Einstein's relations between the stimulated, spontaneous and absorption coefficients, he obtained a system of equations from which the mean number of single quanta united in s-quanta molecules $sh\nu$ in the cavity volume and in the frequency range $d\nu$ is derived as

$$(8\pi\nu^2/c^3)\{\exp[-(sh\nu/kT)]\}\ d\nu. \tag{2.7}$$

The mean total energy per unit volume in the range $d\nu$ is therefore

$$\frac{8\pi h\nu^3}{c^3}\left(\sum_{s=1}^{\infty} \exp[-(sh\nu/kT)]\right) d\nu \tag{2.8}$$

or

$$(8\pi h\nu^3/c^3)\{h\nu/[\exp(h\nu/kT)-1]\}, \tag{2.9}$$

in agreement with Planck's radiation law.

These considerations, which will also be useful later when examining the statistical properties of the radiation field, are useful here to establish another important point. Already, in 1923, when Bothe presented these results, one of the more fundamental properties of stimulated emission was clear, namely that the quantum emitted in the stimulated emission process, besides having exactly the same energy as the stimulating quantum also has the same momentum, i.e. travels in the same direction as the incident quantum. This type of behaviour is exactly what is needed in order to have an amplification process.

For about 30 years, however, the concept of stimulated emission was used only in theoretical works, as will be shown in the following section, and received only marginal attention from the experimental point of view. Even in 1954 the classical monograph by W H Heitler (1904–) on the quantum theory of radiation[12] gives very limited space to this argument whilst dedicating considerable attention to phenomena like resonance fluorescence and Raman scattering.

2.2 The role of stimulated emission in the theory of light dispersion

Einstein's theory of emission and absorption coefficients allowed theoretical physicists to build up a satisfactory quantum theory of scattering, refraction and light dispersion in a few years.

Let us first consider a few elementary definitions relating to the concept of absorption lines. If light from a source emitting a continuum spectrum over some range of frequencies is made parallel and sent through an absorption cell filled with a monatomic gas, the intensity of the transmitted light I_ν may have a frequency distribution as shown in figure 2.1. When this happens the gas is said to have an absorption line at frequency ν_0, ν_0 being the line centre frequency.

The absorption coefficient k_ν of the gas is defined by the relation

$$I_\nu = I_0 \exp(-k_\nu x), \tag{2.10}$$

where x is the thickness of the absorbing layer. From figure 2.1 and equation (2.10), k_ν can be obtained as a function of frequency as shown in figure 2.2. The total width of the curve where k_ν falls to half its maximum value k_{max}, is called the *half linewidth* or *halfwidth*, $\Delta\nu$. In general the absorption coefficient of a gas is given by an expression involving a function of ν and a definite value of

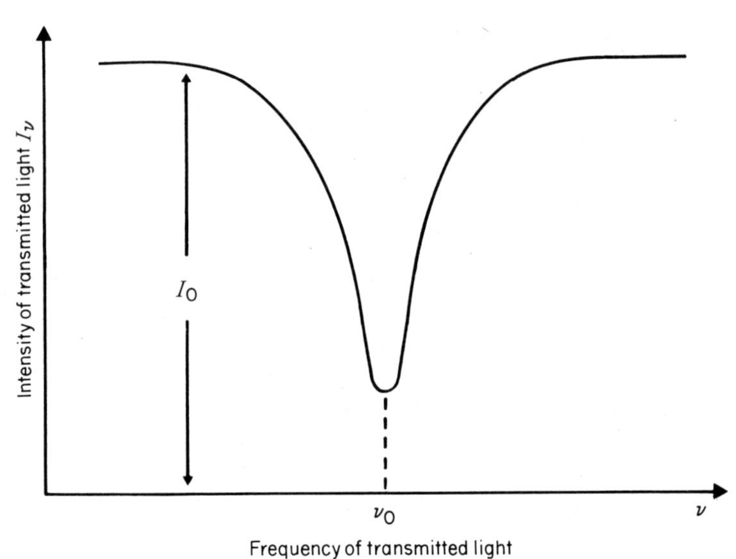

Figure 2.1 Absorption line centred at a frequency ν_0.

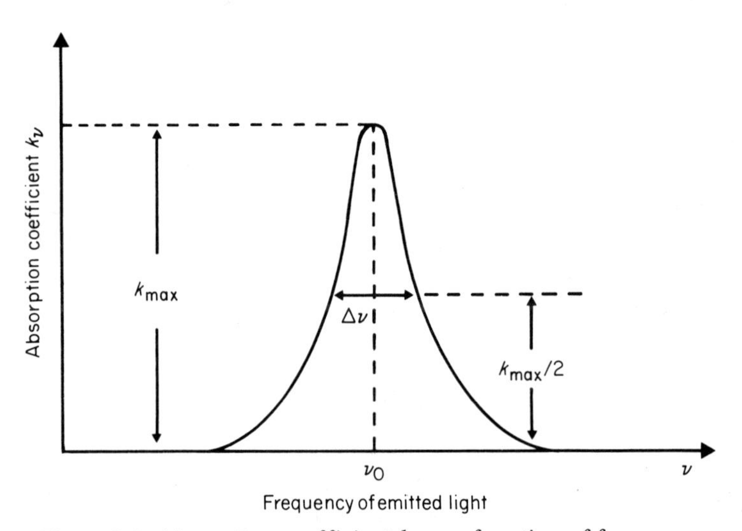

Figure 2.2 Absorption coefficient k_ν as a function of frequency.

k_{max} and $\Delta\nu$, which depend on the nature of the gas molecules, their motion, and the interactions between themselves or with other molecules.

The classical theory of dispersion and absorption of radiation is due principally to P Drude (1863–1906)[13] and W Voigt (1850–1919)[14] and was elucidated fully by H A Lorentz (1853–1928)[15].

In the classical theory the atom is considered to be formed by some oscillators whose frequencies are equal to the absorption frequencies, ν_i. If such oscillators are treated as particles with charge e and mass m, under the action of an oscillating electric field E, we may write for the position vector of the generic oscillator with proper frequency ν_0

$$\ddot{\mathbf{r}} + \gamma\dot{\mathbf{r}} + (2\pi\nu_0)^2\,\mathbf{r} = (e/m)\,\mathbf{E}, \tag{2.11}$$

where γ is a damping factor. For $E = E_0[\exp(2\pi i\nu t)]$ a solution of equation (2.11) is

$$\tau = [(e/m)\,\mathbf{E}]/[4\pi^2(\nu_0^2 - \nu^2) + 2\pi\nu\gamma i], \tag{2.12}$$

and the oscillator has an electric dipole moment

$$\mathbf{p} = e\mathbf{r}. \tag{2.13}$$

The polarisation $\mathbf{P}$ produced by the vibrating electric field is therefore written as

$$\mathbf{P} = eN\mathbf{r} = \alpha\mathbf{E},$$

where N is the number of oscillators (atoms) per unit volume and the polarisability α is the complex number (Drude's formula)

$$\alpha = [(e^2/m)\,N]/[4\pi^2(\nu_0^2 - \nu^2) + 2\pi\nu\gamma i]. \tag{2.14}$$

If ν is not very near to ν_0, the imaginary term in the denominator can be neglected.

In a gas, by introducing a complex refractive index $\hat{n}$ defined by

$$\hat{\epsilon} = 1 + (4\pi P/E) = (\hat{n})^2 = (n = ik)^2, \tag{2.15}$$

where $\hat{\epsilon}$ is the complex dielectric constant and n and k are the real and imaginary parts of $\hat{n}$ respectively, we have therefore

$$(\hat{n})^2 = (n^2 - k^2) - i(2nk) = 1 + 4\pi\alpha. \tag{2.16}$$

By using equation (2.14) we have finally

$$\begin{aligned}(\hat{n})^2 &= (n^2 - k^2) - i(2nk) = 1 + 4\pi[\mathrm{Re}(\alpha) + \mathrm{Im}(\alpha)] \\ &= 1 + 4\pi[\![(e^2N/m)\{(\nu_0^2 - \nu^2)/[4\pi^2(\nu_0^2 - \nu^2)^2 + \nu^2\gamma^2]\} \\ &\quad - i(e^2N/m)\{\nu\gamma/2\pi[4\pi^2(\nu_0^2 - \nu^2)^2 + \nu^2\gamma^2]\}]\!].\end{aligned} \tag{2.17}$$

In the case of a gas of not too high density we can put

$$k \ll 1, \qquad |n - 1| \ll 1, \qquad n^2 - 1 = 2(n - 1),$$

so that we obtain

$$n = 1 + (2\pi e^2N/m)\{(\nu_0^2 - \nu^2)/[4\pi^2(\nu_0^2 - \nu^2)^2 + \nu^2\gamma^2]\}, \tag{2.18}$$

and

$$k = (2\pi e^2 N/m)\{\gamma\nu/2\pi\,[4\pi^2(\nu_0^2 - \nu^2)^2 + \nu^2\gamma^2]\}. \tag{2.19}$$

The behaviour of k and n near the resonant frequency is shown in figures 2.3(*a*) and (*b*).

Far from resonance the refractive index n increases with increasing frequency. The corresponding negative wavelength coefficient $dn/d\lambda$ is called *normal dispersion.*

The first observations of the dispersion phenomena were by Marcus in 1648 and Grimaldi (1613-1703) in 1665[16].

Near resonance, however, the dispersion changes its sign; it is now called *anomalous dispersion* and in this region absorption is also appreciable. In the ideal case of oscillation without damping ($\gamma = 0$) the dispersion, instead of having a maximum and a minimum, tends towards $+\infty$ or $-\infty$ according to whether ν_0 is approached from the lower or higher frequency side, respectively. Figure 2.3(*c*) shows the curve $\epsilon = n^2$ for $\gamma = 0$. P Le Roux (1832-1907) was, in 1862, the first to observe anomalous dispersion[17].

The range where $\epsilon < 0$ is characterised by total reflection. Qualitatively equations (2.18) and (2.19) are in agreement with the measured profiles of

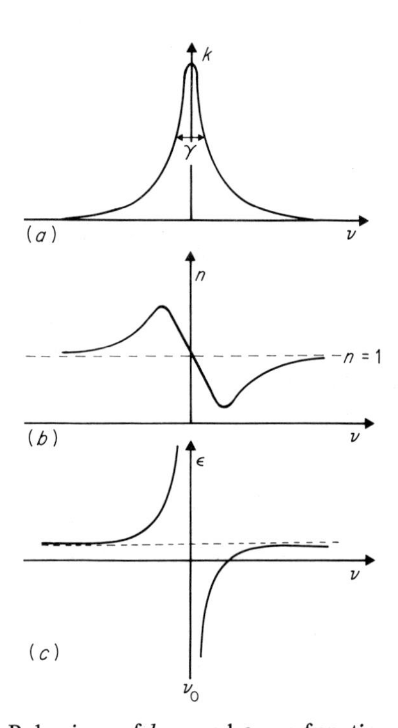

Figure 2.3 Behaviour of k, n and ϵ as a function of frequency.

absorption lines in gases. To have quantitative agreement, however, the concept of effective number of oscillators has to be introduced. Equation (2.18) is accordingly written as

$$n-1=(2\pi e^2/m)\sum_i N_i f_i(\omega_i^2-\omega^2)/[(\omega_i^2-\omega^2)^2+\gamma^2\omega^2], \qquad (2.20)$$

N_i being the number among the N atoms which are in state i.

The factor f_i was interpreted by Drude as the *number* of *dispersion electrons* per atom. W Pauli (1900-1958)[18] called it the *strength* of the oscillator.

The classical equations considered above were in fairly good agreement with experiment and gave a satisfactory interpretation of dispersion and also of absorption when the imaginary part of the refractive index was considered. However, when Bohr's theory of stationary states superseded the classical theory of elastically bound electrons, these formulae, not withstanding their *de facto* validity, lost their theoretical justification completely. The first attempts to formulate a dispersion theory in terms of quantum mechanical concepts, made by P Debye[19] (1884-1966), A Sommerfeld[20] (1868-1951) and C J Davisson[21] (1881-1958) were unsatisfactory, mainly due to the application of the classical perturbation theory to Bohr's atomic model. This application leads to the inevitable conclusion that the Fourier frequencies are also the frequencies of the optical resonance lines – contrary to experience.

The first correct step towards the formulation of the quantum mechanical interpretation of dispersion was taken by Ladenburg. Rudolf Walther Ladenburg plays an important role in our history: as we shall see, he came very close to discovering amplification by stimulated emission. He was born in Kiel, Germany on 6 June 1882 and died in Princeton, New Jersey, on 3 April 1952. He was the third of three sons of the eminent chemist Albert Ladenburg. After school in Breslau, where his father was chemistry professor at the University, Ladenburg went to Heidelberg University in 1900. In 1901 he came back to Breslau and then in 1902 went to Munich, where he took his degree in 1906 with a thesis on viscosity, under Roentgen. From 1906 to 1924 he was at Breslau University first as a *Privat Docent* and from 1909 as Extraordinary Professor.

He married in 1911 and three years later served in the army as a cavalry officer, but later on, during the 1914-18 war, he did research on sound ranging. In 1924, at the invitation of F Haber (1868-1934), he went to the Kaiser Wilhelm Institute at Berlin, the prestigious Institute where Einstein also served, and where he stayed as Head of the Physics Division until 1931 when he went to Princeton.

After the First World War Ladenburg was looking for the theoretical relations connecting the number of oscillators by means of the new method based on Bohr's theory of describing emission and absorption. In 1921 he first introduced the concept of dispersion electrons and gave an expression in terms of the

Einstein A coefficient[22]. He obtained this number by calculating the energy emitted and absorbed by an ensemble of molecules in thermal equilibrium with radiation, on the basis of classical theory on the one hand and of quantum theory on the other, so making an important application of the correspondence principle formulated by Bohr.

To do this he considered $\mathcal{N}$ dispersion electrons per cm^3, able to oscillate freely with frequency ν_1. For a harmonic oscillator of frequency ν_1, if the displacement at the generic time t is $x_0 \cos 2\pi\nu_1 t$, the mean value of the total energy

$$\tfrac{1}{2}m\,[(\mathrm{d}x/\mathrm{d}t)^2 + 4\pi^2\nu_1^2 x^2], \tag{2.21}$$

is

$$\bar{\mathcal{U}} = 2\pi^2 m\nu_1^2 x_0^2. \tag{2.22}$$

The mean energy radiated per second by each such electronic oscillator is[23]

$$(16\pi^4 e^2/3c^3)\,\nu_1^4 x_0^2 \qquad \text{or, therefore,} \qquad (8\pi^2 e^2 \nu_1^2/3mc^3)\bar{\mathcal{U}}\,. \tag{2.23}$$

The energy radiated per second by the $\mathcal{N}$ dispersion electrons is then

$$J_{\mathrm{el}} = (8\pi^2 e^2 \nu_1^2/3mc^3)\,\mathcal{N}\bar{\mathcal{U}}. \tag{2.24}$$

If the molecules are in equilibrium with radiation at temperature T, and if we consider the electrons as spatial oscillators with three degrees of freedom, then between $\bar{\mathcal{U}}$ and the radiation density ρ there exists the relation (Planck)

$$\bar{\mathcal{U}} = (3c^3/8\pi\nu_1^2)\,\rho. \tag{2.25}$$

Therefore,

$$J_{\mathrm{el}} = (\pi e^2/m)\,\mathcal{N}\rho. \tag{2.26}$$

The energy absorbed at equilibrium is of course equal to the radiated energy.

In the quantum theory of Bohr and Einstein the radiation emission from a molecule, as we have seen, is produced in two ways: the spontaneous emission from state k to state i and the stimulated emission. For every transition a quantum of energy $h\nu$ is emitted and therefore the total energy emitted per second is given by

$$J_{\mathrm{Q}} = h\nu_{ik} N_k (A_k^i + B_k^i \rho_{ik}), \tag{2.27}$$

where N_k is the number of molecules in state k. The absorbed energy is

$$A_{\mathrm{Q}} = h\nu_{ik} N_i B_i^k \rho_{ik}, \tag{2.28}$$

where N_i is the number of molecules in state i. At equilibrium these two quantities must be equal and by using Einstein's relation, equation (2.6), between the emission and absorption coefficients and equation (2.3) one readily finds

$$A_{\mathrm{Q}} = J_{\mathrm{Q}} = N_i(c^3/8\pi\nu_{ik}^2)\, A_k^i \rho_{ik}(g_k/g_i), \tag{2.29}$$

and by equating J_{Q} to J_{el} we finally have[24]

$$\mathcal{N} = N_i(mc^3/8\pi^2 e^2 \nu_{ik}^2)\, A_k^i (g_k/g_i). \tag{2.30}$$

Equation (2.30) expresses the constant $\mathcal{N}$ (which can be experimentally deduced from emission, absorption, anomalous dispersion and magnetic rotation measurements, and in the classical theory is interpreted as the dispersion electron number) in terms of quantum quantities N_i and A_k^i. Therefore from measurements, e.g. of anomalous dispersion at different lines of a spectral series, information on the probability of different transitions can be deduced.

Ladenburg applied equation (2.30) to explain his experiments on hydrogen and sodium[22]. In the latter element the oscillator density was about equal to the number of atoms per cm^3. In hydrogen an approximate value of about 4 was found for the ratio of $\mathcal{N}$ relative to the lines H_α and H_β of the Balmer series.

Although Ladenburg does not make explicit mention of it, his work relates to and generalises the classical Drude formula of dispersion in which the atomic frequencies were the absorption frequencies of the atom[25].

Now let us again take equation (2.20), neglect the damping term and substitute $N_i f_i$ by $\mathcal{N}$ as given by equation (2.30). We finally find a formula developed by Ladenburg and Reiche[26]

$$n - 1 = (2\pi e^2/m) \sum_i \{[N_i(mc^3/8\pi^2 e^2 \nu_{ik}^2)\, A_k^i (g_k/g_i)]/(\omega_i^2 - \omega^2)\}. \tag{2.31}$$

This formula is not yet complete. The existence of negative terms in the dispersion formula was not seen before the work by Kramers and Heisenberg.

Hendrik Anthony Kramers was born on 17 December 1894 in Rotterdam, where his father was a physician. He studied at Leyden University, principally with P Ehrenfest (1880–1933), who in 1912 had succeeded H A Lorentz. In 1916 Kramers went to Copenhagen to work with Niels Bohr (1885–1962). When, in 1920, the Bohr Institute of Theoretical Physics opened, Kramers was at first Assistant, and then in 1924, Lecturer. In 1926 he accepted the theoretical physics chair at Utrecht and in 1934 he returned to Leyden as the successor of Ehrenfest, who died in September 1933.

From 1936 until his death on 24 April 1952 Kramers taught at Leyden and payed a number of visits to other countries, including the United States.

During his years at Copenhagen he worked on dispersion problems.

In an early publication[27] he wrote the following expression for the polarisation, P;

$$P = E \sum_i f_i(e^2/m)\,[4\pi^2(\nu_i^2 - \nu^2)]^{-1} \tag{2.32}$$

and observed that a formula of this kind, where the ν_i are equal to the atomic absorption frequencies, represents the experimental results fairly well.

However, the formula does not satisfy the condition, required by the correspondence principle, that, in the region of high quantum numbers, the interaction between the atom and the radiation field tends to coincide with what is expected from classical theory.

To satisfy this condition, Kramers proposed another expression which also contains negative terms corresponding to emission frequencies. To obtain this new expression, Kramers considered the case of an excited atom and proposed to treat it taking into account not only the stationary states i with energy levels higher than the state 1, but also the states j which have lower energy levels than state 1, so that the formula becomes[28]

$$P = (c^3E/32\pi^4)\left(\sum_i A_i^1/[\nu_i^2(\nu_i^2 - \nu^2)] - \sum_j A_1^j/[\nu_j^2(\nu_j^2 - \nu^2)]\right), \tag{2.33}$$

where

$$\nu_j = (E_1 - E_j)/h. \tag{2.34}$$

Equation (2.33) of course relates to a single atom, and a factor has to be adjoined to represent the number of atoms in this state.

In order to derive equation (2.33) Kramers made use of the concept of the 'virtual oscillator', a concept suggested first by J C Slater (1900-1976)[29] and elaborated by Bohr, Kramers and Slater[30] in a celebrated work which suggested that the energy conservation principle could not be valid in elementary processes. Although this idea did not find any immediate confirmation, in the following years the work had a strong influence. It emphasised the notion of virtual oscillators associated with quantum transitions[31].

According to this point of view the dispersion is not to be calculated by considering the real orbit (the stationary state) reacting classically to the exciting wave. Instead, the stationary states appear to be unaffected, except for occasional quantum leaps; so the dispersion is rather to be computed as being due to a set of hypothetical linear oscillators whose frequencies are the spectroscopic ones rather than those of the orbits. Now, as we have seen in classical theory, the atoms behave as electric dipoles of amplitude

$$e^2E/[4\pi^2m(\nu_1^2 - \nu^2)].$$

By considering equation (2.32), we see that according to quantum mechanics the atom behaves with respect to the incident radiation as if it contains a number of linked electrical charges constituting the harmonic oscillators as in classical theory, with each one of these oscillators corresponding to each possible transition between the atomic state and another stationary state.

We may therefore describe the behaviour of a dispersion atom by means of a doubly infinite set (i.e. dependent on two quantum numbers m and n) of virtual harmonic oscillators, with the displacement of the oscillation (m, n) represented by

$$q(m,n) = Q(m,n) \exp[2\pi i \nu(m,n)\, t], \qquad (2.35)$$

where $\nu(m, n)$ indicates the frequency of this oscillator. The set of these virtual harmonic oscillators was called the *virtual orchestra* by A Landé (1888–1975)[32]. The virtual orchestra is then a classical formalism substitution for the radiation and so indirectly it becomes the representation of the quantum radiator itself.

In place of the classical e^2/m we have $c^3A_i^1/8\pi^2\nu_i^2$ for one of the *absorption oscillators*, i.e. the ones corresponding to transitions between state 1 and the higher states, but we have the value $-c^3A_1^j/8\pi^2\nu_j^2$ for one of the *emission oscillators*, i.e. the ones corresponding to transitions between state 1 and the lower states. There is therefore a kind of negative dispersion arising from emission oscillators that can be considered analogous to the *negative absorption* represented by the Einstein B_2^1 coefficient.

In his work[27] Kramers does not say how he derives the dispersion formula. In a second work[33] (which was a reply to a note by G Breit published in the same volume of *Nature*), Kramers gives an account of the derivation. A complete demonstration of the formula is contained in a work[34] written with Heisenberg (1901–1976) who spent the winter 1924–25 in Copenhagen working with Bohr and Kramers. The final form of the main formula of this work was suggested by Heisenberg, but the origin of the work is entirely due to Kramers, as Heisenberg himself stated[35].

In another work Kramers[36] writes:

> If the atom is in one of its higher states, also terms belonging to the second sum inside the brackets of [our equation (2.33)] appear. In the neighbourhood of the frequency ν_{em} of an emission line, the atom will then give rise to an anomalous dispersion of similar kind as in the case of an absorption line, with the difference that the sign of P is reversed. This so called *negative dispersion* is closely connected with the prediction made by Einstein, that the atom for such a frequency will exhibit a *negative absorption*, i.e. light waves of this frequency, passing through a great number of atoms in the state under consideration, will increase in intensity.

A clarification and an extension of the correspondence principle was first given by J H van Vleck in a short paper entitled *A correspondence principle for absorption*[37], and then in more detail in another work[5]. van Vleck's idea is the following; if we wish to calculate the absorption by means of the correspondence principle, we must compare absorption, calculated classically, with the difference between absorption and induced emission, calculated from Einstein's formula. In the limit of large quantum numbers, this difference must be equal to the classical absorption.

In van Vleck's paper the term *induced emission* appears for the first time (§ 3). Moreover he uses freely the expression 'negative absorption' and writes (§ 3):

> The existence of the induced emission term in the quantum theory may at first sight seem strange, but it is well known that it is qualitatively explained in that with the proper phase relations a classical electric wave may receive energy from an atomic system, although on the average (i.e. integrating over all possible phase relations) it contributes more than it receives in exchange. It is therefore the excess of positive absorption over the induced emission which one must expect to find asymptotically (for large quantum numbers) connected to the net absorption in the classical theory.

Still in this period Richard C Tolman (1881–1948) in a paper entitled *Duration of molecules in upper quantum states*[38] writes that '...molecules in the upper quantum state may return to the lower quantum state in such a way as to reinforce the primary beam by "negative absorption".'

Tolman deduced 'from analogy with classical mechanics' that the negative absorption process 'would presumably be of such a nature as to reinforce the primary beam'.

After having so clearly prepared the basis for the invention of the laser, Tolman said that for absorption experiments as usually performed the amount of negative absorption can be neglected.

In the same paper he derived a relation between the integral of the absorption coefficient and the Einstein coefficient $A = 1/\tau$ in an explicit form, although neglecting the effect of negative absorption. A similar relation was already found implicitly in 1920 by Füchtbauer[24]. It was also derived independently by Milne[39] (1896–1950) by means of the following argument and this time taking negative absorption into account.

Let us consider a parallel light beam of frequency in the range between ν and $\nu + d\nu$ and intensity I_ν travelling in the positive x direction through a sheet of atoms limited by two planes at x and $x + dx$. Let us assume that there are N normal atoms per cm^3, δN_ν of which are able to absorb in the frequency range between ν and $\nu + \mathrm{d}\nu$, and N' excited atoms of which $\delta N'_\nu$ are able to emit in this frequency range. By neglecting the effect of spontaneous re-emission and considering that stimulated emission takes place in all directions, the decrease of energy of the beam is given by

$$-\mathrm{d}[I_\nu \delta\nu] = \delta N_\nu \,\mathrm{d}x\,(h\nu/c)\,B_1^2\,(I_\nu/4\pi) - \delta N'_\nu \,\mathrm{d}x\,(h\nu/c)\,B_2^1\,(I_\nu/4\pi), \quad (2.36)$$

where $I_\nu/4\pi$ is the intensity of equivalent isotropic radiation. By rewriting equation (2.36) one has

$$-I_\nu^{-1}(\mathrm{d}I_\nu/\mathrm{d}x)\,\delta\nu = (h\nu/4\pi c)(B_1^2\,\delta N_\nu - B_2^1\,\delta N'_\nu). \quad (2.37)$$

By integrating over the whole absorption line and neglecting the small variation in ν over the line we have (remembering equation (2.10))

$$\int k_\nu \,\mathrm{d}\nu = (h\nu_0/4\pi c)(B_1^2 N - B_2^1 N'), \quad (2.38)$$

where ν_0 is the line centre frequency. By using the relation between Einstein coefficients, we have finally

$$\int k_\nu \,\mathrm{d}\nu = (\lambda_0^2 g_2 N/8\pi g_1 \tau)\,[1 - (g_1/g_2)(N'/N)]. \quad (2.39)$$

This integral was often used to measure τ. The integral $\int k_\nu \,\mathrm{d}\nu$ over the 2537 Å resonance line of Hg was measured by Füchtbauer *et al*[40] in the presence of extrancous gases at pressures between 10 and 50 atm and they found it decreased. An explanation of this in terms of an elevated population in the higher state was given by Mitchell and Zemansky[41].

2.3 Experimental proofs of negative dispersion

It was necessary to wait some years before the first experimental proofs of negative dispersion were found. An expression given by Ladenburg[42] for the dispersion is particularly useful,

$$n - 1 = (e^2/4\pi mc^2)\,[\lambda_{kj}^3/(\lambda - \lambda_{kj})]\,N_j f_{kj}\,[1 - (N_k/N_j)(g_j/g_k)], \quad (2.40)$$

where k and j refer to any two stationary states (k being the higher one) with statistical weights g_k and g_j respectively. N_k and N_j are the numbers of atoms in the two states, λ_{kj} is the wavelength of radiation emitted in the $k \rightarrow j$ transition and

$$f_{kj} = (mc\lambda_{kj}^2/8\pi^2 e^2)(g_k/g_j)\,A_k^j. \quad (2.41)$$

The term $(1 - N_k g_j/N_j g_k)$ was indicated as a *negative dispersion term.* Near resonance the effect of all the other absorption lines becomes negligible and, if the gas excitation is low, so that the negative dispersion term can be taken equal to one, equation (2.40) can be simplified as

$$n - 1 = (e^2 Nf/2\pi mc^2)\,[\lambda_0^3/(\lambda - \lambda_0)], \qquad \lambda_{21} = \lambda_0; \qquad f_{21} = f. \tag{2.42}$$

This is the well known formula of anomalous dispersion used in experiments to obtain precise measurements of f.

The experimental method used was the so-called hook method developed by Roschdestwensky[43]. It was based on the use of a Jamin interferometer as shown in figure 2.4. Tube I can be filled with a gas at known pressure and tube II is kept evacuated. A continuous radiation source is used and the resulting beam is focused on the slit of a spectrograph.

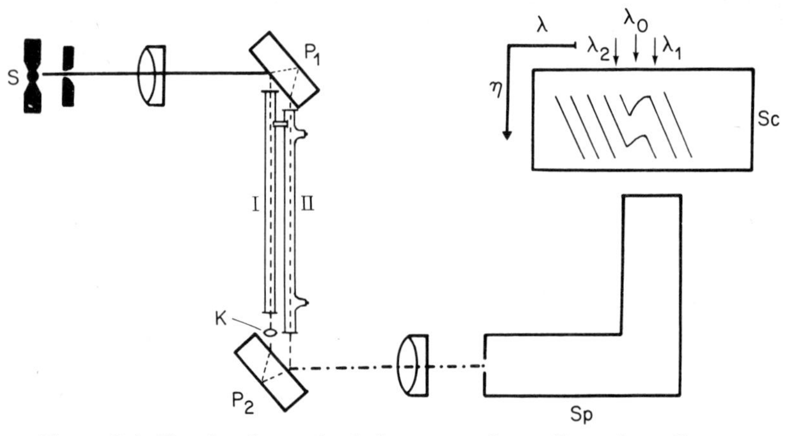

Figure 2.4 The hook method for anomalous dispersion: S, continuum source; I, path through the excited medium; II, path through the normal medium; K, compensator; P_1, P_2, Jamin interferometer plates; Sp, spectrograph; Sc, anomalous dispersion pattern.

With both tubes evacuated and with the compensating plate K removed, the continuous spectrum is crossed by horizontal interference fringes. With the compensating plate in position, the interference fringes are oblique (see figure 2.5, from note 44). If the separation in wavelengths of a convenient number of fringes in the immediate neighbourhood of λ_0 is measured, an important constant, D, of the apparatus can be calculated

$$D = -\lambda_0\,(\text{fringe number/separation in wavelengths of these fringes}). \tag{2.43}$$

If a gas with an absorption line at λ_0 is now introduced into tube I, the oblique interference fringes take the form of a hook symmetrically shaped on both sides of the absorption line.

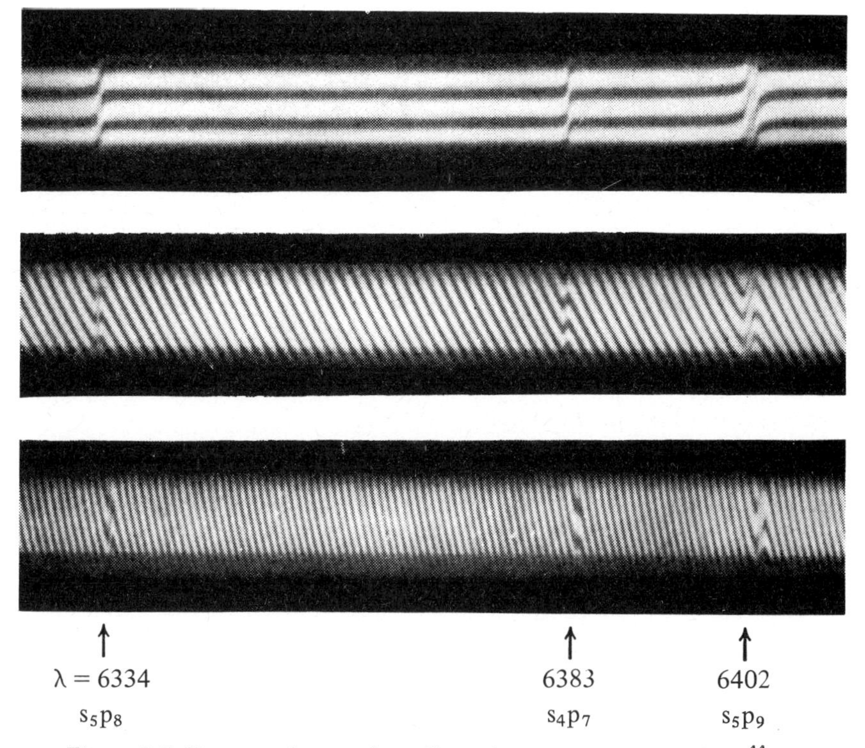

Figure 2.5 Pictures of anomalous dispersion near some neon lines[44].

If A represents the separation in wavelengths of two hooks located symmetrically with respect to the absorption line, then the theory of this method, in connection with equation (2.42) gives

$$f = (\pi mc^2/e^2\lambda_0^3 Nl)\, DA^2, \tag{2.44}$$

from which f can be calculated once N and l (thickness of the column of gas) have been determined.

A simple demonstration of this formula is given in a paper by Ladenburg and Wolfsohn[45].

When the excitation is strong enough, the number of atoms in the higher state N_k can become an appreciable fraction of the number of atoms in the lower state N_j and the expression $(1 - N_k g_j/N_j g_k)$ in equation (2.40) becomes appreciably different from unity. With the hook method, Ladenburg and his collaborators tested the validity of the negative dispersion term during research performed between 1926 and 1930 and published in volumes **48** (1928) and **65** (1930) of *Zeitschrift für Physik*[42, 46, 47]. The most interesting of these studies is the subject of three papers published by Kopfermann and

Ladenburg on the study of dispersion of gaseous neon near the red emission lines[46]. Neon was excited in a tube by means of an electric discharge and dispersion was studied as a function of the discharge current intensity.

The authors excited tubes of 50 and 80 cm length and of 8–10 mm diameter by means of a 20 kV DC generator with currents between 0.1 and 700 mA. The quantity measured was

$$F_{kj} = N_j A^j_k (g_k/g_j)(mc^3/8\pi^2 e^2 \nu^2_{kj})\,[1-(N_k g_j/N_j g_k)], \qquad (2.45)$$

where j is the lower and k the higher level producing the spectral line under investigation, N_j and N_k are the numbers of atoms/cm^3 in these levels, which are of statistical weights g_j and g_k.

At low values of the excitation current up to 60 mA, they found that the quantity $Q = N_k g_j/N_j g_k$ was negligible with respect to unity and the value of F_{kj} for different lines increased with the current (figure 2.6). In figure 2.7 the values of F_{kj} are shown for different lines of figure 2.6 which belong to the same lower level S_5, normalised to their maximum value.

The values so reduced coincide and therefore Ladenburg deduced that the population N_{S_5} of levels S_5 (common to the different lines) changes with current.

The increase of population on level S with the increase of the current was justified through considerations of the atomic excitation and de-excitation mechanisms by invoking the existence of a statistical equilibrium between excited atoms and electrons.

By increasing the current beyond 100 mA, Ladenburg and Kopfermann[48] found a decrease of F. The results of these experiments are shown in figure 2.8, where values of F for different neon lines having the lower level S_5 in common are shown for currents up to 700 mA.

If the values of F of different lines are again reduced to the same scale, these reduced values of F coincide up to 60 mA (*see* Figure 2.9). With higher current values the reduced F values no longer coincide but separate considerably from one another. Those of longest wavelength decrease most, and those of shortest wavelength decrease least, i.e. the smaller the difference of energy between the common lower level S_5 and the different upper states P_k the larger is the decrease of F. This result was correctly interpreted by observing that the greater the current is, the larger is the number of atoms in the upper level P_k. At the same time the number of atoms in the lower state S_5 does not increase, but rather decreases a little above 100 mA. Therefore with increasing current the ratio N_k/N_j increases.

These experiments gave the first experimental proof of the existence of negative terms in the dispersion equation.

Ladenburg ended his paper[44] in 1933 with these words:

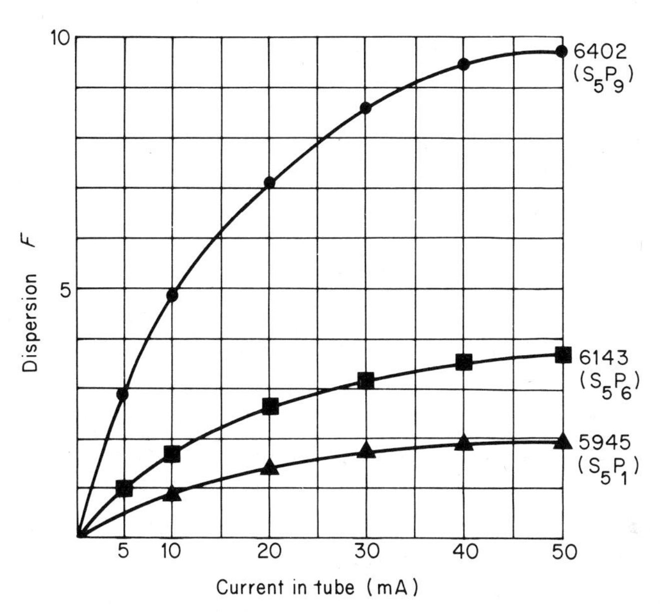

Figure 2.6 Behaviour of F with current in the tube for different S_5 lines of neon[44].

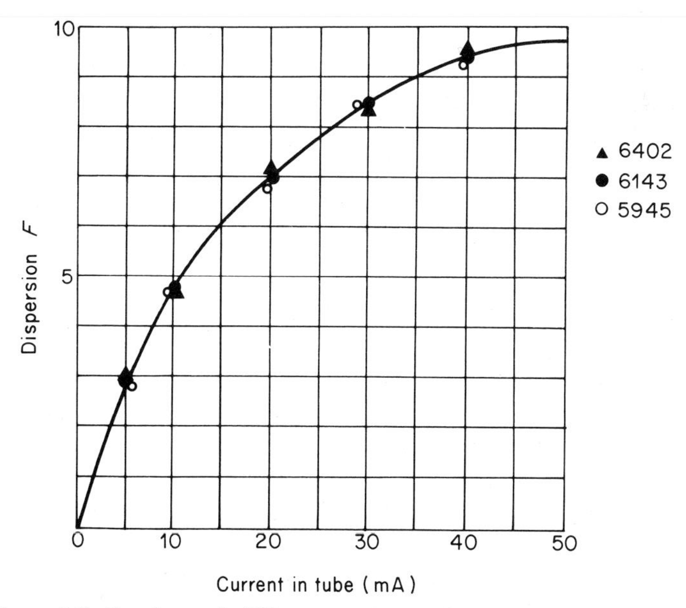

Figure 2.7 F values of different S_5 lines of neon as a function of current, reduced to the same scale[44].

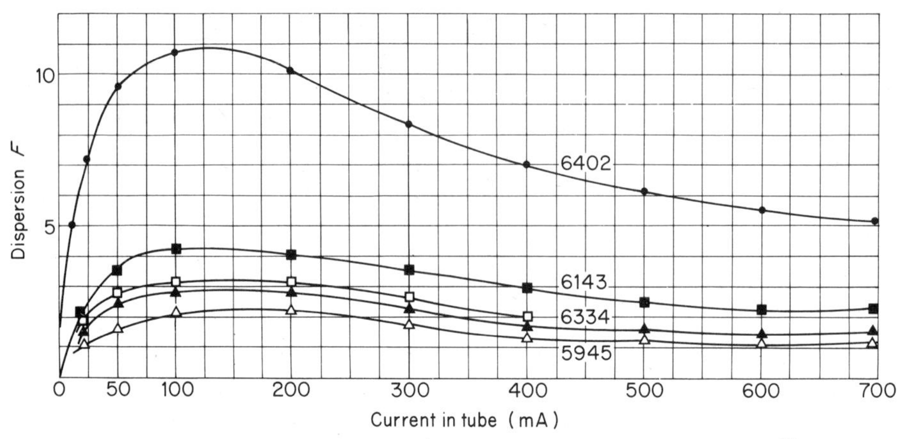

Figure 2.8 Behaviour of F values of S_5 lines with higher currents[44].

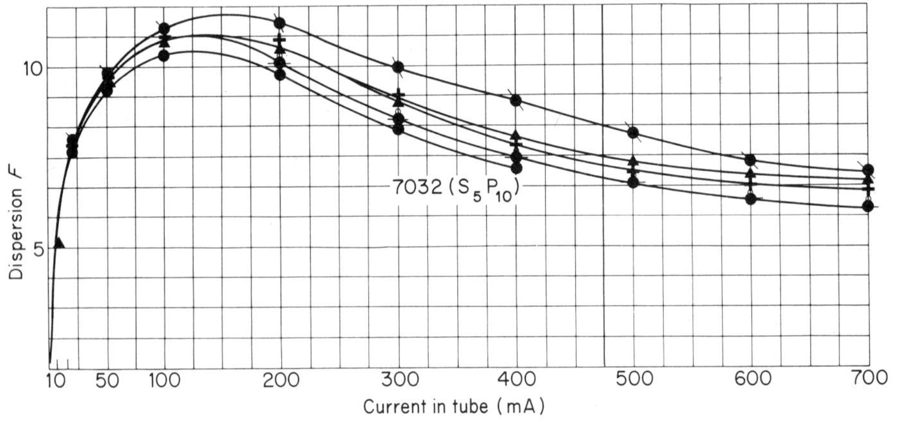

Figure 2.9 F values of figure 2.8 reduced to the same scale[44]. $\lambda=$ ♦, 5945 Å, (S_5P_4); ▲, 6143 Å, (S_5P_6); +, 6334 Å (S_5P_8); -●-, 6402 Å, (S_5P_9); ●, 7032 (S_5P_{10}).

All these different experiments prove without doubt that the population of the P levels does increase with the current, and so much so that above 100 mA, the ratio $Q(=N_k g_j/N_j g_k)$ has appreciable values. Therefore the experiments shown by the curves of [figure 2.9] prove the influence of the negative term in the dispersion formula. This 'negative dispersion' corresponds to the negative absorption of the theory of radiation and to the term -1 in the denominator of Planck's formula for the radiation of a black body as is easily shown by Einstein's derivation of this formula.

If Ladenburg and his co-workers in the experiments with neon had continued by using more intense discharge currents they would have found that the curve persisted in decreasing and became negative.

The curve of anomalous dispersion (figure 2.3) is reversed if the absorption is negative[49]. Schawlow observed in a Conference in Megeve[50] that people did not continue Ladenburg and Kopfermann's studies on anomalous dispersion because they believed so firmly in equilibrium that they thought it was impossible to go so far away from it as to have negative absorption.

2.4 More on negative absorption

In 1940 Fabrikant[51] observed, in his doctoral dissertation:

> For molecular (atomic) amplification it is necessary that N_2/N_1 be greater than g_2/g_1. Such a situation has not yet been observed in a discharge even though such a ratio of populations is in principle attainable.... Under such conditions we would obtain a radiation output greater than the incident radiation and we could speak of a direct experimental demonstration of the existence of negative absorption.

It is interesting to observe that the laser could have been invented, even though 'casually', from 1947 onwards. W E Lamb jr (1913-) and R C Retherford (1912-) were able, by using the methods of radiofrequency spectroscopy, to measure reliably the difference in energy between the $2S_{1/2}$ and $2P_{1/2}$ states of hydrogen. The two states should have been exactly degenerate according to the Dirac equation but a splitting was expected, caused by the coupling of the radiation field with the atom.

The measured splitting (Lamb shift) was first published in 1947[52] and Lamb was awarded the Nobel Prize in 1955 for these researches. In 1950 a subsequent paper on the same subject was published in *Physical Review*[53], where the microwave investigation of the fine structure of hydrogen was fully treated.

In this paper there is an Appendix where the authors analysed the conditions observed in a Wood discharge. They wrote:

> The absorption of radio waves by excited hydrogen atoms in a Wood's discharge tube depends on the population of the various states. These in turn depend on the rates of production and decay of the excited atoms. It would involve a lengthy program of research to make quantitative calculations of these, and we shall be content here with the roughest sort of estimate.

Lamb and Retherford applied their reasoning to the $n = 1$ and $n = 2$ levels of atomic hydrogen, concluding that the effective decay of some of these levels, under particular conditions, was much less than that due to natural lifetime. Consequently the population of these states increased correspondingly, by favouring the occurrence of induced emission when these levels were bathed in radiation.

In particular they considered the transitions induced by radiowaves between the $2^2S_{1/2}$ and $2^2P_{3/2}$ levels of hydrogen, and wrote:

> Let us consider the transitions between $2^2S_{1/2}$ and $2^2P_{3/2}$ induced by radiowaves. If these states are populated in accordance with their statistical weights (equipartition), there will be no appreciable net absorption of RF since the induced emission exactly cancels the induced absorption. (Spontaneous transitions between $2^2P_{3/2}$ and $2^2S_{1/2}$ occur at negligible rate.)
>
> If the population of $2^2S_{1/2}$ is increased relative to $2^2P_{3/2}$, there will be a net absorption of RF. If, on the other hand, $2^2P_{3/2}$ is more highly populated, there will be a net induced emission (negative absorption).
>
> On the basis of the preceding discussion alone, one would expect that the 2p levels would be about five to ten times more populated than the 2s levels. In that case, one would expect to find a negative absorption and as estimated below, a large one.

Some calculation on the attenuation coefficient followed and they concluded,

> In view, however, of the extreme crudeness of the numerical estimates, it is possible that some appreciable departure from equipartition may exist, and that an absorption or induced emission could be detected. It is therefore highly desirable that a search for such effects should be made, especially under discharge conditions which do not favour equipartition.

Lamb wrote later[54]:

> The concept of negative absorption was new to us at the time, and we were unaware of the earlier references... I think that we understood that the radiation would be coherent, as was the input signal. However, we did not associate negative absorption with self-sustained oscillation.
>
> Even if we had done so, at least three factors would have kept us from inventing the maser: (1) our interest was centred on the fine structure of hydrogen, (2) the smallness of the expected absorption (gain), and doubt as to its sign, and (3) the ready availability of oscillators at the frequency used.

Notes

1 A Einstein, *Mitt. Phys. Ges., Zurich* **16** no. 18, 47 (1916). This work was subsequently published in *Z. Phys.* **18**, 121 (1917) and is translated into English in the book *Sources of Quantum Mechanics* edited by B L van der Waerden (North-Holland: Amsterdam, 1967) and in *The Old Quantum Theory* (Pergamon: Oxford and New York, 1967) edited by D ter Haar. See also A S Eddington *Phil. Mag.* **1**, 803 (1925), L Ornstein and F Zernike, *Versl. Akad. Amsterdam* **28**, 280 (1919)

2 W Wien, *Verh. Dtsch Phys. Ges.* **18**, no. 13/14, 318 (1916)

3 This expression had already been introduced in a previous work by Einstein, *Verh. Dtsch Phys. Ges.* **16**, 820 (1914). See also Einstein's work on specific heat in *Ann. Phys., Lpz.* (4) **22**, 180 (1907)

4 Actually if one averages over all the phases, the classical theory would not have predicted stimulated emission. Einstein either paid no attention to this or understood it was not important. See also W Heitler, *The Quantum Theory of Radiation 2nd edn* (Oxford University Press: Oxford, 1944) chapter 1 section 5

5 J H van Vleck, *Phys. Rev.* **24**, 330 (1924)

6 According to relation (2.6) of the text, spontaneous emission increases with ν^3 and this behaviour reflects the circumstance that excited atoms can decay spontaneously in any of the $8\pi\nu^2\,\mathrm{d}\nu/c^3$ modes per unit volume, while stimulated emission must occur in the single mode of the incident photon. Relation (2.6) can therefore be written

$$A_m^n\,\mathrm{d}\nu = (8\pi\nu^2\,\mathrm{d}\nu/c^3)\,h\nu B_m^n,$$

which tells us that the spontaneous transition probability is equal to the product of the number of modes and the absorption rate for a quantum per mode. In other words, the spontaneous emission probability per mode is equal to that of (negative or positive) absorption in the presence of a quantum in that mode.

7 Einstein's formulae were extended to the case of non-sharp energy levels by R Becker, *Z. Phys.* **27**, 173 (1924) and to the interaction laws between radiation and free electrons by A Einstein and P Ehrenfest, *Z. Phys.* **19**, 301 (1923)

8 L S Ornstein and H C Burger, *Z. Phys.* **24**, 41 (1924)

9 However, it is well known that the light emitted by an atom resembles a spherical wave in many of its properties. This, for example, is the case in the discussion of many interference and scattering phenomena. Einstein's argument about unidirectional emission at first sight seems in conflict with this view. The argument was discussed in the following years and comes under close examination in a paper by G Breit (*Rev. Mod. Phys.* **5**, (1933), § 3). The reconciliation of the two demands is found in Heisenberg's uncertainty principle. In Einstein's discussion the momentum of the light quantum and the momentum of the atom are definitely known. According to the uncertainty principle this automatically excludes knowledge of the positions of the atoms and thus makes a discussion of interference impossible.

On the other hand, in discussing interference we suppose the position of the atoms to be known so that the momentum cannot be ascertained and under these circumstances the atom may be said to emit spherical waves.

If the atom were held fixed it would emit a spherical wave. Breit shows that, due to recoil, the spherical waves emitted at each point of its recoil trajectory by the atom under integration over the recoiling time interfere and give rise to unidirectional quanta.

10 M Jammer, *The Conceptual Development of Quantum Mechanics* (McGraw Hill: New York, 1966) chapters I and IV

11 W Bothe, *Z. Phys.* **20**, 145 (1923)

12 W H Heitler, *The Quantum Theory of Radiation, 3rd edn* (Oxford University Press: Oxford, 1954)

13 P Drude, *Ann. Phys., Lpz.* (4) **1**, 437 (1900)

14 W Voigt, *Magneto-Elektrooptik* (G B Teubner: Leipzig, 1908) p103

15 H A Lorentz, *Theory of Electrons* (G B Teubner: Leipzig, 1916)

16 See S A Korff and G Breit, *Rev. Mod. Phys.* **4**, 471 (1932)

17 F P Leroux, *C.R. Acad. Sci., Páris* **40**, 126 (1862)

18 W Pauli, *Quantentheorie, Handbuch der Physik* vol 23 (Springer: Berlin, 1926) p 87

19 P Debye, *Munchener Berichte* (1915) pp1–26

20 A Sommerfeld, *Ann. Phys., Lpz.* **53**, 497 (1917)

21 C Davisson, *Phys. Rev.* **8**, 20 (1916)

22 R Ladenburg, *Z. Phys.* **4**, 451 (1921). This work is translated into English in *Sources of Quantum Mechanics* edited by B L van der Waerden (North-Holland: Amsterdam, 1967)

23 Expressions (2.23) were simple applications of the well known Larmor formula; J Larmor, *Phil. Mag.* **44**, 503 (1897)

24 See also Ch Füchtbauer, *Phys. Z.* **21**, 322 (1920)

25 Compare on this point van der Waerden's comment '... it follows that Drude's formula [equation (2.14) here] is valid for a set of classical harmonic oscillators. Ladenburg does not write out his derivation, but it was generally known at his time, and we may safely assume that he had such a derivation in mind. Hence we may say that Ladenburg replaced the atom, as far as its interaction with the radiation field is concerned, by a set of harmonic oscillators with frequencies equal to the absorption frequencies ν_i of the atom.

This idea is not explicitly formulated in Ladenburg's paper, but it is implicitly contained in it, and Ladenburg's contemporaries realised this...' [in van der Waerden, *Sources of Quantum Mechanics* (North-Holland: Amsterdam, 1967) p11].

26 R Ladenburg and F Reiche, *Naturwiss.* **11**, 584 (1923)

27 H A Kramers, *Nature* **113**, 673 (1924)

28 The formula given by Kramers contained an additional factor 3; this was a consequence of his hypothesis that free oscillations were parallel to the incident field, while equation (2.33) in the text here assumes that all atomic orientations are equiprobable. See also the reference in note (5).

29 J C Slater, *Nature* **113**, 307 (1924)

30 N Bohr, H A Kramers and J C Slater, *Phil. Mag.* **47**, 785 (1924)

31 A discussion of the importance of this work can be found for example in M Jammer, *The Conceptual Development of Quantum Mechanics* (McGraw-Hill: New York, 1966) section 4.3

32 A Landé, *Naturwiss.* **14**, 455 (1926)

33 H A Kramers, *Nature* **114**, 310 (1924)

34 H A Kramers and W Heisenberg, *Z. Phys.* **31**, 681 (1925). An English translation is available in *Sources of Quantum Mechanics* edited by B L van der Waerden (North-Holland: Amsterdam, 1967)

35 In *Sources of Quantum Mechanics* edited by B L van der Waerden (North-Holland: Amsterdam, 1967) p16

36 H A Kramers, *Skand. Mat. Kongr.* (1925) 143, reproduced in H A Kramers *Collected Scientific Papers* (North-Holland: Amsterdam, 1956) p 321

37 J H van Vleck, *J. Opt. Soc. Am.* **9**, 27 (1924)

38 R C Tolman, *Phys. Rev.* **23**, 693 (1924)

39 E A Milne, *Mon. Not. R. Astron. Soc.* **85**, 117 (1924)

40 C Füchtbauer, G Joos and O Dinkelacker, *Ann. Phys., Lpz.* **71**, 204 (1923)

41 See A C G Mitchell and M W Zemansky, *Resonance Radiation and Excited Atoms* (Cambridge University Press: Cambridge, 1934) pp113–4

42 R Ladenburg, *Phys. Z.* **48**, 15 (1928)

43 D Roschdestwensky, *Ann. Phys., Lpz.* **39**, 307 (1928) and *Trans. Opt. Inst., Leningrad* **2**, no. 13 (1921). See also J Jamin, *Ann. Chem. Phys.* **52**, 163 (1858); L Puccianti, *Nuovo. Cim.* **2**, 257 (1901); R Ladenburg and St Loria, *Z. Phys.* **9**, 875 (1908)

44 R Ladenburg, *Rev. Mod. Phys.* **5**, 243 (1933)

45 R Ladenburg and G Wolfsohn, *Z. Phys.* **63**, 616 (1930)

46 R Ladenburg and H Kopfermann, *Z. Phys.* **48**, 26 and 51 (1928) and *Z. Phys.* **65**, 167 (1930)

47 R Ladenburg and S. Levy, *Z. Phys.* **65**, 189 (1930); A Carst and R Ladenburg, *Z. Phys.* **48**, 192 (1928). All these works were synthesised in the beautiful paper by Ladenburg (note 44).

48 See note 46 and H Kopfermann and R Ladenburg, *Z. Phys. Chem.* **139**, 378 (1928)

49 Compare A Kastler, *Ann. Phys., Paris* **7**, 57 (1962)

50 A Schawlow in a discussion after a paper at the Second International Conference on Laser Spectroscopy, Megeve, 23–7 June 1975

51 V A Fabrikant, *Thesis* (1940) quoted in F A Butayeva and V A Fabrikant *Investigations in Experimental and Theoretical Physics, A Memorial to S G Landsberg* (USSR Academy of Science Publications: Moscow, 1959) pp62–70

52 W E Lamb jr and R C Retherford, *Phys. Rev.* **72**, 241 (1947)

53 W E Lamb jr and R C Retherford, *Phys. Rev.* **79**, 546 (1950)

54 W E Lamb jr, *Physical concepts of the developments of the maser and laser* in *Impact of Basic Research on Technology* edited by B Kursunoglu and A Perlmutter (Plenum Press: New York, 1973)

3

Intermezzo: Magnetic resonance and optical pumping

3.1 Introduction

There is no immediate explanation of why more than 20 years had to elapse before the invention of masers and lasers, notwithstanding the fact that the concepts of stimulated emission and negative absorption had been well established since the 1930s. One may argue that one of the reasons for this may be that until the 1950s efforts to produce coherent radiation were directed essentially towards radiowaves, which are, in practice, always emitted coherently.

The concept of coherence, moreover, was not yet fully understood; nor was the connection with stimulated emission completely appreciated.

We may also observe that the 20 years between the 1930s and the 1950s were not lost in so far that an acquisition of knowledge took place which was later used in masers and lasers. Moreover these efforts were directed towards problems which, although they intrinsically contained ideas basic to the making of stimulated emission devices, had, however, a completely different end in mind. Amongst the arguments then under consideration were studies on magnetic resonance and optical pumping. During World War II the main efforts were directed towards the production and detection of microwaves for radars. The technical problems which were solved, amongst others, were the development of high-power microwave generators, called magnetrons, to produce the radar signal; the construction of sensitive crystal detectors to detect the echo; the development of electronic methods of distinguishing the echo over the background noise; and the perfection of narrow-band amplifiers, lock-in detectors and other noise-reducing circuits to increase the sensitivity of the radar system.

New fields of science – such as microwave and semiconductor device engineering – grew, and the resultant new techniques were invaluable in the development of electronic and nuclear resonance.

Magnetic resonance opened the way both to the understanding of many concepts and to the use of many techniques which were later to be used in masers and lasers. The parallel development of spectroscopy finally led to the optical detection of magnetic resonance and to the technique of optical pumping.

However, in the subsequent development of masers and lasers, magnetic resonance and optical pumping had noticeably different weights. The studies of magnetic resonance led to a consideration of the possibility of changing the population of the various energy levels and of introducing population inversion through the concept of negative temperature, so leading the way to new approaches and methodologies which had, as a natural output, the maser principles. Optical pumping, on the other hand, although clearly showing the way to extend the same concepts to the optical domain, had little influence on the later developments in this direction and lasers came out more as a natural extension of masers to shorter wavelengths rather than as a possibility offered by optical pumping.

Much later, A Kastler, the inventor of optical pumping, who was awarded the Nobel Prize for the work in 1966, said himself that he and his collaborators had 'never worked on induced emission problems which are at the base of laser operation'[1].

3.2 Magnetic resonance

Magnetic resonance involves the reorientation of a magnetic dipole as a whole in an external field, or the reorientation of one of the magnetic dipoles existing in an atom or a nucleus with respect to the others. In the first case, transitions are produced between Zeeman components of an energy level; in the second, the transitions are among fine-structure or hyperfine components. To observe the effect, two magnetic fields are needed: one static, which removes degeneracy by splitting the energy levels, and the other oscillating, to induce transitions between two states. In this way absorption (or emission) of radiation takes place which produces changes in the equilibrium distribution of energy levels.

The phenomenon is in some way analogous to the electric dipole transition case, but it is much more involved. In the electric case the levels between which transitions take place (i.e. the oscillating electrical dipole in the classical representation) always exist, being the energy levels of electrons in the atom. In the magnetic case, magnetic energy levels must first be created by some suitable external field and, in general, their spacing is proportional to this field. The study of these magnetic phenomena is also complicated by the circumstance that paramagnetic susceptibilities are lower than the electrical ones by several orders of magnitude. Moreover, the theory of paramagnetism was developed rather late[2].

The problem of how fast the average magnetic moment in a paramagnetic substance responds to a sudden change in the magnetic field in which the substance is placed had already been faced in the 1920s by W Lenz[3], P E Ehrenfest[4] and G Breit (1899-) and H Kamerlingh Onnes[5,6] (1853-1926) before the electronic and magnetic resonance techniques were developed. In the 1930s the problem of how a magnetic system reaches thermal equilibrium received great attention and this interest was stimulated by the first experiments on adiabatic demagnetisation and magnetic relaxation.

E Majorana (1906-1938)[7] and I I Rabi (1898-)[8] discussed theoretically the magnetic resonance absorption and Rabi and his collaborators[9] detected magnetic resonance in atomic beams in which transitions between energy levels corresponding to different nuclear spin orientations in a strong constant magnetic field were induced by a radio-frequency magnetic field[10]. For these experiments Rabi was awarded the Nobel Prize in 1944.

The same method was also successfully applied by Alvarez and Bloch in 1940[11] to the measurement of the magnetic moment of the neutron.

In condensed matter Waller (1898-)[12] in a paper, now famous, which appeared in 1932, had already distinguished the two main relaxation phenomena, spin-spin and spin-lattice relaxation.

The important concept of spin-lattice relaxation was later taken up again in thermodynamic theory by Casimir and Dupré[13], in which the magnetic crystal was considered as being divided into two systems, each possessing its own temperature. One system contains the magnetic degrees of freedom and goes into internal equilibrium at a temperature T_M in a very short time (the spin-spin relaxation time $t_2 \simeq 10^{-10}$ s). The other system, the lattice or phonon system, contains all the other degrees of freedom and is at a temperature T_L which may be different from T_M. The time needed to establish thermal equilibrium between the two systems is the spin-lattice relaxation time τ_{SL} or t_1 (typically of the order of milliseconds). The nuclear spin systems have spin-lattice relaxation times enormously greater than the electronic spin-lattice relaxation times[14].

Excitation of magnetic levels was at that time done in order to study atomic levels and nuclear spins. Gorter (1907-1980) had considered in the late 1930s and early 1940s[15,16] the possiblity that nuclear magnetic moment precession in an external field could give rise to macroscopic effects. In 1936 he attempted to detect nuclear resonance in solids by observing an increase in temperature, and he showed remarkable insight by attributing the negative results of his experiment to a long spin-lattice relaxation time ($t_1 > 10^{-2}$ s). Later on he tried again, this time by measuring magnetic dispersion; but again he had no result. (The early history and development of the field of paramagnetic relaxation studies until the end of the Second World War has been reviewed in a monograph by Gorter[17], where he gives a detailed account of theoretical and experimental results known to him at that time.)

3.3 Bloch, Purcell and Zavoisky

The first successful experiments to detect magnetic resonance by electromagnetic effects were carried out independently by F Bloch at Stanford[18], E M Purcell at Harvard[19] and E Zavoisky in the USSR[20].

Bloch and Purcell were awarded the Nobel Prize in 1952 for their researches. They introduced magnetic resonance through two different paths which were, however, substantially similar. Zavoisky was the first to observe transitions between fine-structure levels of the fundamental state in paramagnetic salts (paramagnetic electronic resonance).

Felix Bloch was born in Zurich, Switzerland, on 23 October 1905. He entered the Federal Institute of Technology (Eidgenössische Technische Hochschule) in Zurich in 1924. After one year's study of engineering he decided instead to study physics and changed to the Division of Mathematics and Physics at the same institution. During the following years he studied under Professors Debye, Scherrer, Weyl and Schrödinger. He was interested initially in theoretical physics. After Schrödinger left Zurich in the autumn of 1927 he continued his studies with Heisenberg at the University of Leipzig, where he received his degree of Doctor of Philosophy in the summer of 1928 with a dissertation dealing with the quantum mechanics of electrons in crystals and the theory of metallic conduction. In the year that followed he worked with Pauli, Kramers, Heisenberg, Bohr and Fermi.

After Hitler's ascent to power, Bloch left Germany in 1933. A year later he accepted a position at Stanford University, California. There he started experimental research. In 1936 he published a paper in which he showed that the magnetic moment of free neutrons could be measured through the observation of scattering in iron, and showed that in this way polarised neutrons could be obtained. During the war he was also engaged in the early stages of the work on atomic energy at Stanford University and Los Alamos and later in counter measures against radar at Harvard University. Through this latter work he became acquainted with the modern developments in electronics which, towards the end of the war, suggested to him, in conjunction with his earlier work on the magnetic moment of the neutron, a new approach to the investigation of nuclear moments in solids. In 1945, immediately after his return to Stanford, he began the study of *nuclear induction*, as he was later to call it.

His contribution to the theory of magnetism in matter is not restricted to this[21]: he also held important scientific positions. In 1954 he was the first Director General of CERN in Geneva, the large European organisation for high-energy research.

Edward Mills Purcell was born in Taylorville, Illinois on 30 August 1912. In 1929 he entered Purdue University in Indiana when he graduated in electrical engineering in 1933. His interest had already turned to physics and K Lark-

Horovitz, the great professor to whom solid-state physics in the USA is so indebted, allowed him to take part in experimental research in electron diffraction. After one year spent in Germany at the Technische Hochschule, Karlsruhe, where he studied under Professor W Weizel, he entered Harvard University, and received a PhD in 1938. After serving two years as instructor in physics at Harvard he joined the Radiation Laboratory at MIT, which was established in 1940 for military research and the development of microwave radar. He became Head of the Fundamental Developments Group in the Radiation Laboratory which was concerned with the exploration of new frequency bands and the development of new microwave techniques. The discovery of *nuclear resonance absorption*, as he called it, was made just after the end of the war and at about that time Purcell returned to Harvard as Associate Professor of Physics. He became Professor of Physics in 1949.

Eugenii Konstanovich Zavoisky was born in Kazan in 1907 into a doctor's family. He studied and then worked at Kazan University. He was interested almost from his student days in the use of radio-frequency electromagnetic fields for the study of the structure and properties of matter. Commencing in 1933 he performed exploratory experiments on the resonant absorption of radio-frequency fields by liquids and gases. In 1941 he became the first to use the modulation of a constant magnetic field by an audio-frequency field in such experiments. In 1944 he discovered electron paramagnetic resonance, which became the subject of his doctoral dissertation. During the years 1945-7 he performed a series of important experiments, recording paramagnetic dispersion curves in the resonance range and obtaining electron paramagnetic resonance in manganese. Later on he became associated with the Kurchatov Institute of Atomic Energy in Moscow where he worked for more than 20 years.

He made contributions to various fields of nuclear physics – developing, amongst other things, the scintillation-track chamber in 1952; and to plasma physics – he discovered magneto-acoustic resonance in 1958. He was awarded the Lenin and State Prizes. He died in 1976. His studies became known in the West only after the Second World War.

The announcement of the first experiments on magnetic resonance was given independently by Bloch and Purcell within a month of each other. In the January 1946 issue of the *Physical Review*, E M Purcell, H C Torrey and R V Pound (1919-)[19] in a short letter to the editor, received on 24 December 1945, announced that they had observed absorption of radio-frequency energy, due to transitions induced between energy levels which corresponded to different orientations of the proton spin in a constant applied magnetic field in a solid material (paraffin). In this case there are two levels the separation of which corresponds, in a field of 7100 Oe, to a frequency, ν, of 29.8 MHz.

They observed:

> Although the difference in population of the two levels is very slight at room temperature ($h\nu/kT \simeq 10^{-5}$), the number of nuclei taking part is so large that a measurable effect is to be expected providing thermal equilibrium can be established... A crucial question concerns the time required for the establishment of thermal equilibrium between spins and lattice. A difference in the population of the two levels is a prerequisite for the observed absorption, because of the relation between absorption and stimulated emission. Moreover, unless the relaxation time is very short the absorption of energy from the radio-frequency field will equalise the population of the levels more or less rapidly, depending on the strength of this RF field.

The experimental arrangement consisted of a resonant cavity adjusted to resonate at about 30 MHz. The inductive part of the cavity was filled with paraffin and the resonator was placed in the gap of the large cosmic-ray magnet in the Research Laboratory of Physics at Harvard. Radio-frequency power was introduced into the cavity at an extremely low level (10^{-11} W) with its magnetic field everywhere perpendicular to the steady field. When the strong magnetic field was varied slowly an extremely sharp resonance absorption was observed (figure 3.1, originally figure 1 from reference 22).

In the next issue of the *Physical Review*, again in letters to the editor, there appeared a short note by F Bloch, W W Hansen (1909–1949) and M Packard (1921–)[23] which had been received on 29 January 1946. In it was written:

> The nuclear magnetic moments of a substance in a constant magnetic field would be expected to give rise to a small paramagnetic polarisation, provided thermal equilibrium be established, or at least approached. By superposing on the constant field (z direction) an oscillating magnetic field in the x direction, the polarisation, originally parallel to the constant field, will be forced to precess about that field with a latitude which decreases as the frequency of the oscillating field approaches the Larmor frequency. For frequencies near this magnetic resonance frequency one can, therefore, expect an oscillating induced voltage in a pick-up coil with axis parallel to the y direction. Simple calculation shows that with reasonable apparatus dimensions the signal power from the pick-up coil will be substantially larger than the thermal noise power in a practicable frequency band.
>
> We have established this new effect using water at room temperature and observing the signal induced in a coil by the rotation of the proton moments.

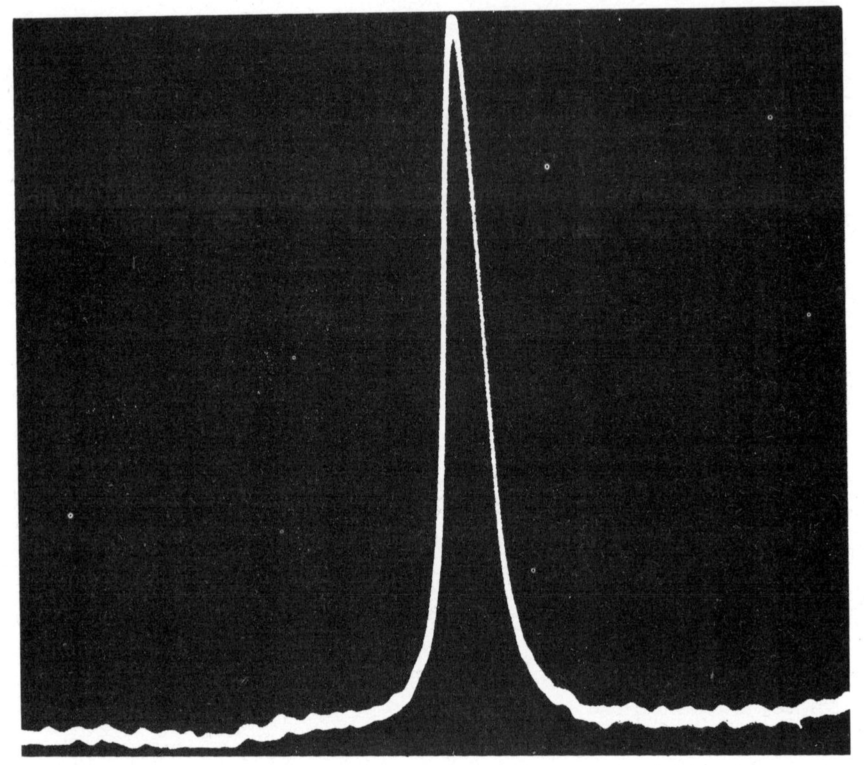

Figure 3.1 Proton resonance (absorption) curve in ferric nitrate solution as derived from note 22. From Bloembergen *et al*, *Phys. Rev.* **73**, 679 (1948); © The American Physical Society.

Therefore at resonance the macroscopic magnetisation vector can be turned out of the z direction. The resulting transverse magnetisation was detected by an induction coil placed normal to both fixed and alternating fields. With Bloch's method the sign of nuclear magnetic moments can be determined, it being possible to observe the sense of rotation of the Larmor precession of the nuclear magnets.

Purcell and Bloch's experiments used nuclear spin levels. Zavoisky's experiment was carried out using electron magnetic levels. The essential difference is that, in the case of electron spin resonance, the applied external magnetic fields are acting on the magnetic moment associated with the electron, which in general comes from the contribution of its spin and orbital angular motion, and the incoming electromagnetic radiation induces transitions between energy states of this electron.

3.4 Bloch equations

The theoretical explanation of the Bloch experiment was given shortly afterwards in a famous work entitled *Nuclear Induction*, published in *Physical Review*[24], in which Bloch gave a phenomenological explanation of magnetic resonance by taking into account relaxation effects and using an entirely classical treatment.

He considered a paramagnetic sample in a magnetic field H. The equation of motion for the macroscopic magnetisation vector **M**, describes the precession of **M** around **H**

$$\mathrm{d}\mathbf{M}/\mathrm{d}t = \gamma(\mathbf{M} \times \mathbf{H}), \tag{3.1}$$

where γ is the gyromagnetic ratio. In this equation the interactions among spins and between spins and the lattice are not considered. The precession frequency of **M** around **H**, i.e. the Larmor frequency, is given by[25]

$$\omega_{\mathrm{L}} = \gamma H. \tag{3.2}$$

Without interaction there is no change in the component of **M** along the direction of **H**, which we shall take parallel to the z axis.

Bloch then considered what happens if the magnetisation is not in equilibrium and described phenomenologically the relaxation of **M** towards equilibrium through simple exponential laws with two chracteristic times[26]. One time, t_2, describes how fast the transverse components M_x or M_y die out, and the other one, t_1, describes how fast the component along z attains the equilibrium value

$$M_{z_0} = \chi_0 H. \tag{3.3}$$

The resulting phenomenological equations were written

$$\begin{aligned} \mathrm{d}M_{x,y}/\mathrm{d}t &= \gamma(\mathbf{M} \times \mathbf{H})_{x,y} - M_{x,y}/t_2, \\ \mathrm{d}M_z/\mathrm{d}t &= \gamma(\mathbf{M} \times \mathbf{H})_z - (M_z - M_{z_0})/t_1, \end{aligned} \tag{3.4}$$

which are usually referrred as to Bloch equations. Then Bloch considered the existence of a weak oscillating radio-frequency field along the x direction of the kind

$$H_x = 2H_1 \cos \omega t. \tag{3.5}$$

The solution of equations (3.4) was found by replacing this field with a rotating field around the z direction[27]

$$H_x = H_1 \cos \omega t; \qquad H_y = \mp H_1 \sin \omega t, \tag{3.6}$$

with the sign of H_y, and therefore the sense of rotation, being negative or positive, depending upon whether the sign of γ is positive or negative. By calling the x and y components of the magnetisation vector in the rotating system u and v respectively, we have

$$M_x = u \cos \omega t - v \sin \omega t \tag{3.7a}$$

$$M_y = v \cos \omega t + u \sin \omega t. \tag{3.7b}$$

The introduction of rotating coordinates[28] is equivalent to replacing the magnetic field H with an effective field of constant direction equal to

$$H_{\text{eff}} = [H_0 + (\omega/\gamma)]\,\mathbf{k} + H_1\,\mathbf{i}, \tag{3.8}$$

where **i** and **k** are unit vectors along the x and z axes, respectively, of the rotating system. The angle θ between H_{eff} and H_0, which goes from 0 to π is given by

$$\tan\theta = H_1/[H_0 + (\omega/\gamma)] = \omega_1/(\omega_0 - \omega),$$

with $\omega_1 = -\gamma H_1$. In this rotating system the motion of the magnetic moment M is a Larmor precession around the effective field H_{eff}, with angular velocity

$$-\gamma H_{\text{eff}} = -\gamma\{[H_0 \mp (\omega/\gamma)]^2 + H_1^2\}^{1/2}. \tag{3.9}$$

Bloch introduced the set of abbreviations

$$\omega_0 = |\gamma|H_0; \qquad \omega_1 = |\gamma|H_1; \qquad \beta = 1/\omega_1 T_2;$$

$$\alpha = 1/\omega_1 T_1; \qquad \delta = (\omega_0 - \omega)/\omega_1; \qquad \tau = \omega_1 t.$$

In terms of these quantities equations (3.4) in the rotating system become

$$\begin{aligned} du/d\tau + \beta u + \delta v &= 0 \\ dv/d\tau + \beta v - \delta u + M_z &= 0 \\ dM_z/d\tau + \alpha M_z - v &= \alpha M_0. \end{aligned} \tag{3.10}$$

Through these equations Bloch was able to deal with both his nuclear induction and the absorption experiment of Purcell. These experiments were both done by making the field H_0 along z constant and changing the frequency of the field H_1, passing through the resonance, or else by making the frequency constant and causing H_0 to vary around the resonance value. These variations were considered by Bloch in the two limiting cases in which the passage through resonance takes place in a time much shorter than the relaxation times involved (*adiabatic fast passage*) or in the opposite case in which the passage time is much longer than these times (*slow passage*).

In the first case the variation of δ is slow, and both quantities α and β are assumed to be small compared to unity. The first condition implies

$$|d\delta/d\tau| \ll 1.$$

In order to have $\alpha \ll 1$, $\beta \ll 1$ it is necessary that either the relaxation times t_1 and t_2 are sufficiently large or that the amplitude $2H_1$ of the oscillating field is sufficiently large.

With the three quantities $|d\delta/d\tau|$, α and β small compared to unity a particular solution can be written in the convenient form

$$M_x = [M/(1+\delta^2)^{1/2}] \cos \omega t \tag{3.11a}$$

$$M_y = \mp [M/(1+\delta^2)^{1/2}] \sin \omega t \tag{3.11b}$$

$$M_z = M\delta/(1+\delta^2)^{1/2}. \tag{3.11c}$$

The quantity M depends in a rather involved manner on the nuclear relaxation times. Under favourable conditions it may be expected to be of the order of the equilibrium polarisation M_0. While its absolute value $|M|$ still represents the instantaneous magnitude of the polarisation the quantity M itself is not necessarily positive but may have both signs, depending upon the positive or negative values which δ had assumed in the past. The amplitude of M_y is therefore proportional to $1/(1+\delta^2)^{1/2}$.

The opposite limiting case is that of 'slow passage' through resonance or short relaxation times. In this case Bloch found for arbitrary values of α, β

$$u = \{(|\gamma| H_1 t_2^2 \Delta\omega)/[1 + (t_2\Delta\omega)^2 + (\gamma H_1)^2 t_1 t_2]\} M_0$$

$$v = -\{(|\gamma| H_1 t_2)/[1 + (t_2\Delta\omega)^2 + (\gamma H_1)^2 t_1 t_2]\} M_0$$

$$M_z = \{[1 + (t_2\Delta\omega)^2]/[1 + (t_2\Delta\omega)^2 + (\gamma H_1)^2 t_1 t_2]\} M_0, \tag{3.12}$$

where

$$|\gamma| H_0 - \omega = \Delta\omega.$$

For nuclear induction experiments it is evidently favourable to have u as large as possible.

In this case all three components of polarisation vanish at resonance. The amplitudes of M_x and M_y are proportional to the function [neglecting $(1/\gamma H_2 T_2)^2$],

$$f(\delta) = \delta/(\delta^2 + t_1/t_2). \tag{3.13}$$

Both functions $1/(1+\delta^2)^{1/2}$ and $f(\delta)$ are shown in figure 3.2[29].

From equation (3.11b) and from the definition of M it is evident that the magnitude of the signal induced by the M_y component of nuclear polarisation depends not only on M, but also, in a rather involved way, on the relaxation times and the magnitude and variation in velocity of δ. In the special case of rapid passage an estimate of the induced RF voltage is obtained by considering that a receiver coil with N turns around a cross sectional area A of the sample

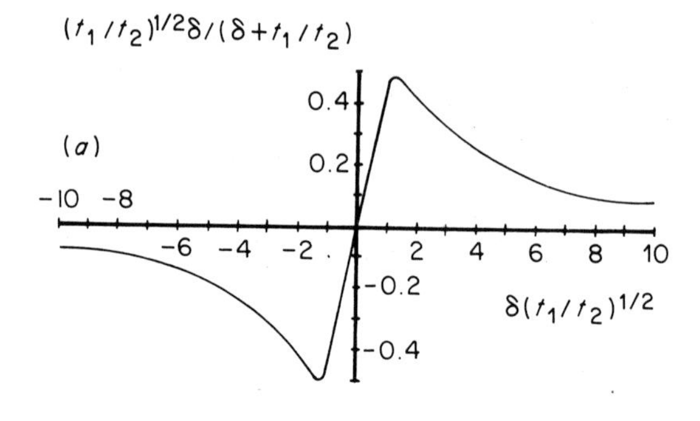

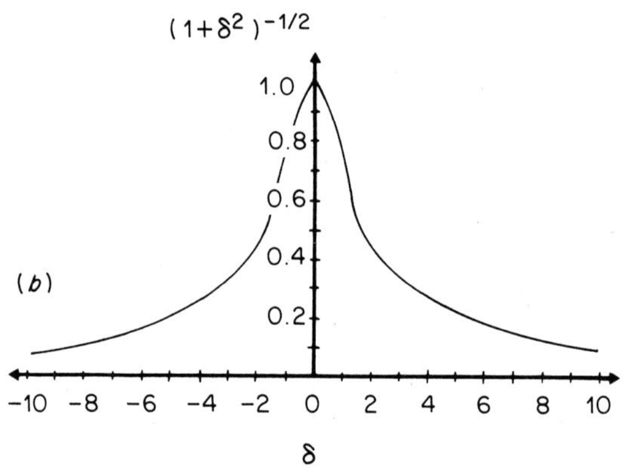

Figure 3.2 Plot (a) of the function $f(\delta)$ in normalised form and, (b) of the function $1/(1+\delta^2)^{1/2}$ as a function of δ.

has a magnetic flux through it given by

$$\Phi(B) = NAB_y = NA4\pi M_y = \mp 4\pi NAM \sin \omega t/(1+\delta^2)^{1/2}. \qquad (3.14)$$

The induced voltage V across the terminals of the coil is

$$V = -(1/c)(\mathrm{d}\Phi/\mathrm{d}t) = \pm(4\pi/c)\, NAM\omega \cos \omega t/(1+\delta^2)^{1/2}, \qquad (3.15)$$

where the variation of δ has been considered sufficiently slow that its time derivative can be neglected when compared to that of $\cos \omega t$.

Finally it can be shown immediately that an expression for magnetic susceptibility can be derived from equation (3.12). In fact, from the relation

$$M_x = \chi H_x = (\chi' - \mathrm{i}\chi'')\, H_x, \qquad (3.16)$$

where χ' and χ'' are the real (connected to dispersion) and imaginary (connected to absorption) parts of χ, respectively, we have

$$\chi' = u/H_1 \qquad \chi'' = -v/H_1. \qquad (3.17)$$

The real and imaginary parts of χ have, therefore, the same behaviour as the functions u and v and are qualitatively still represented by the curves of figure 3.2.

The Bloch and Purcell experiments allow the observation either of χ' and χ'' separately or of various combinations of these quantities.

These different possibilities had been discussed earlier by Bloch[24]. He had noted in this paper that in order to observe nuclear induction it was evidently advisable to have u as large as possible.

Its maximum is obtained for

$$\Delta\omega = (1/t_2)[1 + (\gamma H_1)^2 t_1 t_2]^{1/2}, \tag{3.18}$$

and has the value

$$u_{\max} = \{|\gamma| H_1 t_2 / [1 + (\gamma H_1)^2 t_1 t_2]^{1/2}\} M_0. \tag{3.19}$$

This value again increases monotonically with H_1 and, for

$$H_1 \gg 1/[|\gamma|(t_1 t_2)^{1/2}], \tag{3.20}$$

becomes

$$u_{\max\max} = (t_2/t_1)^{1/2} M_0. \tag{3.21}$$

To obtain maximum absorption it is necessary, on the other hand, to make v as large as possible, since it is this quantity which, through equation (3.7*a*) determines the out-of-phase part of M_x. v has its maximum

$$v_{\max} = -\{(|\gamma| H_1 t_2)/[1 + (\gamma H_1)^2 t_1 t_2]\} M_0, \tag{3.22}$$

for $\Delta\omega = 0$. Unlike $u_{\max}$ this quantity does not increase monotonically for increasing H_1, but decreases for large values of H_1. This phenomenon is today known as saturation. The best possible choice is

$$H_1 = 1/[|\gamma|(t_1 t_2)^{1/2}], \tag{3.23}$$

which yields

$$v_{\max\max} = (t_2/t_1)^{1/2} M_0. \tag{3.24}$$

Magnetic resonance soon became a field of its own and as such is treated in several excellent textbooks.

3.5 Experimental proofs of population inversion

Bloch undertook his experiments on nuclear induction with the collaboration of W W Hansen and M Packard[30]. This followed the theoretical treatment[24] in the same issue of *Physical Review*. One of the experiments described is of particular interest to us. After having determined that the relaxation time of the substance with which they were working (water) was between $\frac{1}{2}$ s and 1 min, they did the following experiment to determine its value more precisely[30].

> Starting at a time t_1, with H_{dc} held for a considerable previous time above the resonance field H^* a positive signal was observed on the right-hand side of the oscillogram ... as presented on trace a of figure 8 [our figure 3.3]. Thereupon the field was quickly (i.e., during about one second) lowered to a value sufficiently below resonance to make the signal appear on the left-hand side of the oscillogram and then was held fixed at this new value.
>
> As was expected ... the signal was originally still positive. However, during the following few seconds it was observed to decrease in magnitude, then to disappear and therefore to grow again with negative values until after several seconds it had reached its full negative value ... as presented on trace c of figure 8 [our figure 3.3]. This extraordinary reversal of the signal under fixed external conditions represents actually a direct visual observation of the gradual adjustment of the proton spin orientation to the changed situation caused by the previous change of the magnet current. The fact that it takes place during a time interval of several seconds evidently indicates the relaxation time, likewise, to be of the order of a few seconds.

During the inversion time of the signal the spin population was inverted, but apparently Bloch paid no attention to this, concentrating instead on the problems of the determination of the relaxation time, its exact meaning and its value. The inversion of population obtained in this way (adiabatic fast passage) was later used (1958) to create population inversions in two-level solid-state masers (*see* Chapter 4).

The next year N Bloembergen, a young Dutch physicist, whom we shall discuss in detail later, together with Purcell and Pound, published a paper[22] where important considerations on relaxation times were developed and calculations presented on the absorption lineshape in an experiment of nuclear magnetic resonance absorption. This paper also included, for the first time, a discussion in terms of populations of the various magnetic levels, which were to prove fundamental to the description of the behaviour of masers and lasers. The authors considered a substance containing N_0 cm^{-3} nuclei, of spin I and magnetic moment μ, placed in a strong uniform magnetic field H_0 along the z axis, and subjected to a weak oscillating field $H_x = 2H_1 \exp(2\pi i \nu t)$, $H_y = 0$. The probability of a single transition in which an m_i changes to an m_i' can be found with the aid of the standard formula for magnetic dipole transitions:

$$W_{m_i \to m_i'} = (8\pi^3/3h^2)|\langle m_i|M|m_i'\rangle|^2 \rho_\nu. \qquad (3.25)$$

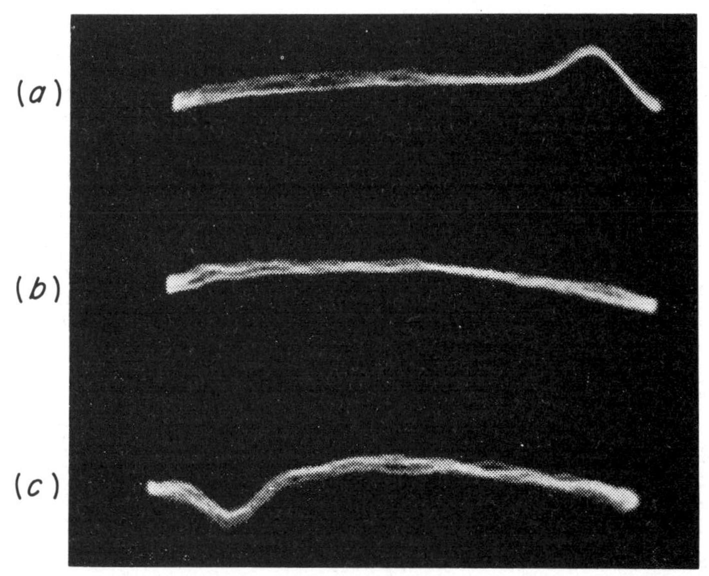

Figure 3.3 Photographic record of the proton signal from water. The three traces from top to bottom correspond to an AC modulation of the magnetic field superimposed upon a DC value which in (*a*) is above, (*b*) at and (*c*) below the resonant field H^*. From Bloch *et al*, *Phys. Rev.* **70**, 474 (1946); © The American Physical Society.

M is the magnetic moment operator. Ordinarily ρ_ν represents the energy density, in unit frequency range, in the isotropic unpolarised radiation field. We have to deal here with radiation of a single frequency from levels of a finite width, which we describe by the observed shape of the absorption line, $g(\nu)$. The shape function $g(\nu)$ is to be normalised so that

$$\int_0^\infty g(\nu)\,\mathrm{d}\nu = 1. \tag{3.26}$$

The radiation field in this case consists simply of an oscillating magnetic field of single polarisation. The equivalent isotropic unpolarised radiation density is calculated as

$$\rho_\nu = 3H_1^2 g(\nu)/4\pi. \tag{3.27}$$

One has finally

$$\begin{aligned} W_{m\to m-1} &= (\pi/3)\,\gamma^2(I+m)(I-m+1)\,\rho_\nu \\ &= \tfrac{1}{4}\gamma^2 H_1^2 g(\nu)(I+m)(I-m+1), \end{aligned} \tag{3.28}$$

where m is the magnetic quantum number.

Equation (3.28) gives the probability for a transition $m \to m-1$, involving the absorption from the radiation field of the energy $h\nu = h\gamma H_0/2\pi$. If the spin system is initially in equilibrium at the temperature T, the population of each level m exceeds that of the next higher level, $m-1$, by

$$N_m - N_{m-1} \simeq [N_0/(2I+1)](h\nu/kT). \tag{3.29}$$

The approximation (3.29) is an extremely good one, for, in the cases considered $h\nu/kT \simeq 10^{-6}$. The net rate at which energy is absorbed is now

$$P_a = [N_0/(2I+1)][(h\nu)^2/kT] \sum_{m=I}^{-I+1} W_{m\to m-1}$$

$$= \gamma^2 H_1^2 N_0 (h\nu)^2 I(I+1)\, g(\nu)/6kT. \tag{3.30}$$

This is also a calculation of the imaginary part of the magnetic susceptibility χ'', being

$$P_a = 4\pi\chi''\nu H_1^2. \tag{3.31}$$

The expression so derived is exact, provided that the original distribution of population amongst the levels remains substantially unaltered. If the strength of the field H_1 increases we may expect to have a redistribution of populations in the various levels.

Bloembergen, Purcell and Pound applied this reasoning to the case $I=\frac{1}{2}$ so that they had to deal with two levels.

Let n denote the surplus population of the lower level: $n = N_{+1/2} - N_{-1/2}$, and let n_0 be the value of n corresponding to thermal equilibrium at the lattice temperature. In the absence of the radio-frequency field, the tendency of the spin system to come to thermal equilibrium with its surroundings was described by an equation of the form

$$\mathrm{d}n/\mathrm{d}t = (1/t_1)(n_0 - n), \tag{3.32}$$

where the characteristic time t_1 is the spin-lattice relaxation time.

The presence of the radiation field requires the addition to equation (3.32) of another term,

$$\mathrm{d}n/\mathrm{d}t = (1/t_1)(n_0 - n) - 2nW_{1/2\to -1/2}. \tag{3.33}$$

A steady state is reached when $\mathrm{d}n/\mathrm{d}t = 0$, or, using equation (3.28), when

$$n/n_0 = [1 + \tfrac{1}{2}\gamma^2 H_1^2 t_1 g(\nu)]^{-1}. \tag{3.34}$$

If the maximum value of $g(\nu)$ is expressed in terms of a quantity t_2^* defined by

$$t_2^* = \tfrac{1}{2} g(\nu)_{\max}, \tag{3.35}$$

the maximum steady-state susceptibility in the presence of the RF field is thus reduced, relative to its normal value, by the 'saturation factor',

$$[1 + \gamma^2 H_1^2 t_1 t_2^*]^{-1}.$$

The quantity t_2^* defined by equation (3.35) is a measure of the inverse line-width.

Effects due to having more particles in excited states were later discussed by W E Lamb[31], as we saw in Chapter 2. In the same year, Pound[32] showed the possibility of having a variation in population of levels saturated in an RF field.

The next year (1951) Purcell and Pound, in a very short note in *Physical Review*[33] entitled *A Nuclear Spin System at Negative Temperature*, introduced the concept of negative temperature and showed the existence of a negative absorption.

They considered a nuclear absorption experiment and reasoned in the following manner. At field strengths which allow the system to be described by its net magnetic moment and angular momentum, a sufficiently rapid reversal of the direction of the magnetic field should result in a magnetisation opposed to the new sense of the field. The reversal must occur in such a way that the time spent below a minimum effective field is so small compared with the period of the Larmor precession that the system cannot follow the change adiabatically.

They found that in a LiF crystal a zero field resonance occurred at about 50 kHz and the relaxation time was rather long. Therefore they put the sample in a magnetic field and, after equilibrium was reached, suddenly inverted the direction of the magnetic field. The inversion time was made shorter than the spin-lattice relaxation time, and so the configuration of nuclear spins had no time to change during the field inversion.

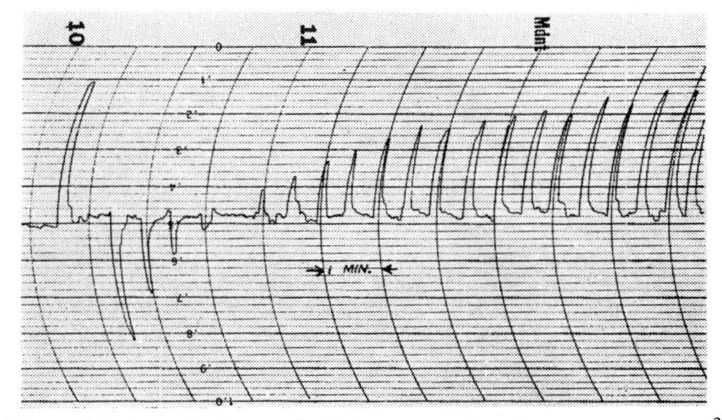

Figure 3.4 A typical record of the reversed nuclear magnetisation[33]. From Purcell and Pound, *Phys. Rev.* **81**, 279 (1951); © The American Physical Society.

During the short time in which spins stayed inverted a negative absorption (i.e. an emission) occurred.

This effect is shown in figure 3.4[33] which is one of the records obtained by sweeping the impressed frequency periodically back and forth through the resonance frequency. The peak at the extreme left is the normal resonance curve, before the field is reversed. Just to the right of this sweep the field has been reversed and the next resonance peak is seen to point downwards, corresponding to negative absorption. The negative peaks get weaker until finally the state is reached where the positive and negative absorption cancel out because there is then equal population of the upper and lower states. The gradually increasing positive peaks show the reestablishment of the thermodynamic equilibrium population.

3.6 The concept of negative temperature

We have already observed how Casimir and Dupré[13] considered that in the phenomenon of spin-lattice relaxation it might be convenient to describe the spin system as being a separate system from the lattice, each of the two systems being itself in thermodynamic equilibrium, but not reaching equilibrium with the other system until later. In this case the spin system can be described by its separate temperature (*spin temperature*). The probability that a given spin has some energy E is therefore given as

$$p(E) = \exp(-E/kT),$$

where T can be taken as the temperature of the system. The concept of *spin temperature* was discussed later by several authors[34]. The distribution function for a system with two temperatures is, for example, described by

$$\exp[(\mathscr{H}_1/kT_1) - (\mathscr{H}_2/kT_2)],$$

where $\mathscr{H}_i$ is the Hamiltonian function of the ith system described by its temperature, T_i; this holds as long as the two systems remain separate from each other.

Purcell and Pound pointed out that the population in the Zeeman levels immediately after the field reversal in their experiments could still be described by the usual Boltzmann distribution if a negative temperature $-T_0$ was considered. In their very short note they did not give any detailed account of this new concept, so a few remarks are given here. Negative temperature is only possible for a system whose energy levels have an upper bound. It allows population inversions to be described. If we consider the Boltzmann distribution law between two levels, 1 and 2,

$$N_2 = N_1 \exp[-(E_2 - E_1)/kT], \qquad (3.36)$$

this law can be used at any instant where T is the so-called *spin temperature*. Equation (3.36) can be considered as defining an instantaneous temperature, T_s, of the spin system in terms of the instantaneous populations N_1 and N_2.

At thermal equilibrium this spin temperature equals the ambient temperature T. Using it, a population inversion $N_2 > N_1$ is then described by a negative temperature.

The ratio

$$\Delta N/N = (N_1 - N_2)/N = \exp[-(E_2 - E_1)/kT] - 1,$$

is shown in figure 3.5 as a function of $\Delta E/kT = (E_2 - E_1)/kT$. At very low temperatures, $T_s \to +0$, all the spins are frozen in the lower level and $\Delta N \simeq N_1$ because there is insufficient thermal energy present to lift any spin into level 2. As the spin temperature increases, more spins are thermally excited into the upper level until, as $T_s \to +\infty$, the populations of the two levels approach equality, $N_1 \simeq N_2$ and $\Delta N \to +0$.

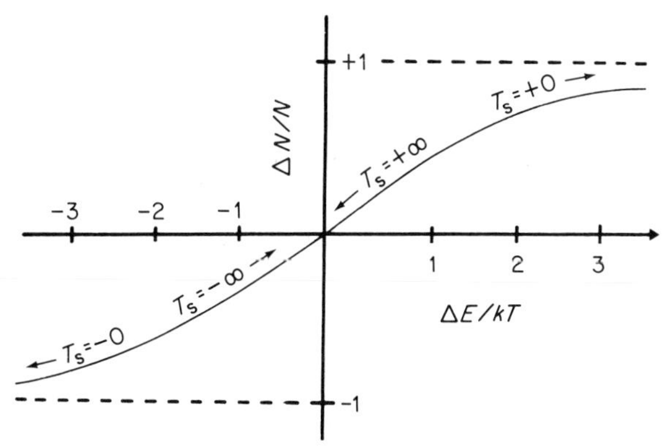

Figure 3.5 Relative population difference $\Delta N/N = (N_1 - N_2)/N$ as a function of $\Delta E/kT$.

When the population difference becomes negative, (inversion) $N_2 > N_1$ and this is possible only if T becomes a negative number. From figure 3.5 it is apparent that the condition $T = +\infty$ passes over continuously into the condition $T = -\infty$, leading to the observation that negative temperatures are essentially 'hotter', not 'colder', than positive temperatures. As one increases the spin temperature, passing through $T_s = +\infty$ to $T_s = -\infty$ and on towards $T_s = -0$, the population of the upper level continuously increases, until at $T_s = -0$ all the spins are 'heated' into the upper level and $\Delta N = -N_2$. It is clear that only systems with an upper limit to their energy spectrum can have negative temperatures.

A complete explanation of the meaning of negative temperature was given seven years after its first introduction in an experiment by Abragam and

Proctor[35] where they demonstrated the identities of both spin temperature and thermodynamic temperature.

A comment by Purcell, reported by N Bloembergen[36] on these experiments, was 'it is like receiving a marriage licence seven years after the child is born'.

Coming back to the Purcell and Pound experiment, a particularly striking feature is that, regardless of how the LiF sample is oriented when it is placed back in the strong field, the spins always remember to point in the 'wrong' direction, or in other words, the negative temperature is always achieved, showing clearly that the negative temperatures have a physical reality. This is due to the circumstances that immediately after the reversal of the field the spin–spin and Zeeman temperatures are not equal, but one is in fact the negative of the other. If they were to have remained equal during the reversal of the field, the experiment would have been equivalent to adiabatic demagnetisation down to $H=0$ and then remagnetisation with the field applied in the opposite direction. The resulting temperature would then always be positive, and the spins would always manage to point in the 'right' direction.

Purcell and Pound's paper gave only the barest outline of the experiments. The concept of negative temperature was subsequently treated and completed by a number of researchers[37]. The purely philosophical or conceptual aspects were discussed in a later paper by Ramsey[38].

In a paper which appeared in 1952, N Ramsey[39] had already written:

> Pound, Purcell and Ramsey† performed a series of experiments with LiF crystals which have a very long relaxation time. They found, among other things, that the spin system is essentially isolated for times which vary from 15 s to 5 min and that, for times short compared to these, the spin system can be placed in a state of negative temperature. In a negative temperature state the high-energy levels are occupied more fully than the low, and the system has the characteristic that, when radiation is applied to it, stimulated emission exceeds absorption.

In the Purcell and Pound experiment the signal they observed was produced by the decay of the inverted population in Zeeman levels. Nobody paid any attention to this method which allowed inversion to be obtained, nor to the fact that systems at negative temperature, when in connection with a microwave cavity or a waveguide, could give coherent amplification through the stimulated emission processes. Probably this was due to the circumstance that the inversion method used gave only transient population inversions.

The method of adiabatic magnetisation eventually used to obtain population inversions useful for building up paramagnetic masers was not proposed until

† R V Pound, *Phys. Rev.* **81**, 156 (1951); E M Purcell and R V Pound, *Phys. Rev.* **81**, 279 (1951); N F Ramsey and R V Pound, *Phys. Rev.* **81**, 278 (1951).

much later, by Strandberg[40] in 1956 and by Townes and his collaborators[41], who did not then succeed in making it work. The method was, however, made to work two years later[42] (see Chapter 4).

3.7 The Overhauser effect

In the autumn of 1951 Al Overhauser came to the University of Illinois, Urbana, as a postdoctoral student from Berkeley, where he had just completed his PhD thesis with Charles Kittel. For his thesis Overhauser had calculated the spin-lattice relaxation time of conduction electrons in metals[43]. No one had actually observed the electron spin resonance of conduction electrons at that time.

Subsequent to completing his thesis, Overhauser had noted a striking result contained in his calculation of the contribution of nuclear spins to relaxing the conduction electron spins: if the conduction electron spin populations were equalised (e.g. by saturating their electron spin resonance) the population difference of the nuclei would be greatly enhanced.

He already had this idea by the time he arrived at Illinois and discussed a possible experiment with C P Slichter, then a young Assistant Professor there. The first step in the experiment was to detect the conduction electron spin resonance, and Slichter proposed this argument as a thesis to one of his students, Don Holcomb. They searched unsuccessfully for several months in all of the good metals such as copper, silver, gold and also in sodium. Eventually Don Holcomb switched to studying the nuclear resonance in lithium metal and they gave up.

Then in the 15 November issue of *Physical Review*, Griswold, Kip and Kittel[44] announced their discovery of the conduction electron spin resonance at microwave frequencies in lithium and sodium.

Slichter immediately repeated the experiment, and with Tom Carver, who had joined Slichter's group in the meantime, immediately set to work to try to verify Overhauser's nuclear polarisation prediction.

At about this time Overhauser made his first public announcement of his polarisation scheme in a 10 minute contributed talk at the Washington meeting of the American Physical Society in April 1953[45]. Present at the talk were, amongst others, Purcell, Rabi, Ramsey and Bloch all of whom later entered into a vigorous discussion with Slichter who had been mentioned by Overhauser as attempting with T Carver to verify his scheme. A watcher said later to him 'You got a Nobel grilling'![46].

Overhauser predicted that if the spin resonance of conduction electrons in a metal is saturated this should increase by a factor of several thousand the nuclear polarisation for metals in which the nuclei reach thermal equilibrium with the lattice by means of the magnetic hyperfine interaction with the conduction electrons. The explanation of this rather surprising result involved solving the problem of the interaction between the electron spin magnetic moment and the nuclear spin magnetic moment, as we shall see later. A full discussion appeared in the 15 October issue of *Physical Review*[47]. Overhauser's proposal was agreed with scepticism and there was a belief that his scheme violated the second law of thermodynamics[48].

In the Overhauser effect one deals with a paramagnetic metal with nuclear spin, usually a feebly paramagnetic alkali, in which the paramagnetism arises from the conduction electrons. A strong constant magnetic field H is applied, whose direction we take as the z axis. The electrons are exposed to a microwave field of a frequency which is resonant to the value of H employed, and which is powerful enough to give a considerable degree of saturation. Overhauser proposed a method which involved observing the shift of the ESR frequency brought about by the polarisation of the nuclei[49].

In a few months Thomas R Carver and Charles P Slichter eventually got the experimental proof. They looked at the strength of the nuclear resonance absorption, which is proportional to the population difference between adjacent nuclear Zeeman levels, by observing the enhancement of the nuclear resonance in metallic lithium produced by electron saturation[50, 51]. The experiment was performed in a static magnetic field of 30.3 G. The sample containing small pieces of lithium dispersed in oil was placed in the tank coil of a 50 W oscillator operating at 84 MHz, the Larmor frequency for the electrons in the magnetic field. The nuclear resonance was observed using a 50 kHz crystal controlled oscillator on an oscilloscope. Figure 3.6 summarises the results. The top line shows the appearance of the ordinary lithium nuclear resonance, which is so weak at the considered frequencies as to be completely lost in noise. The second line was photographed after the electron saturating oscillator was turned on and the Li resonance now appears strongly. For comparison, the proton line in glycerin (also at 50 kHz) is shown in the bottom line.

The result of this experiment which strikingly confirmed Overhauser's theory gave the opposite behaviour to that which one would naively expect. Saturation tends to equalise the population of the upper and lower Zeeman states, and so produces a higher effective temperature. One might anticipate that the nuclear resonance is correspondingly weakened, but in fact the reverse is found.

The explanation of this paradox is that the electron and nuclear Zeeman temperatures are not the same, and that when one is raised, the other is lowered.

This is a case in which systems at different temperatures coexist when the thermal contacts between them are weak. The explanation of the effect can be found by considering that, in metals, the direct spin-lattice relaxation time for the nuclear spins is extremely long. By far the most dominant relaxation

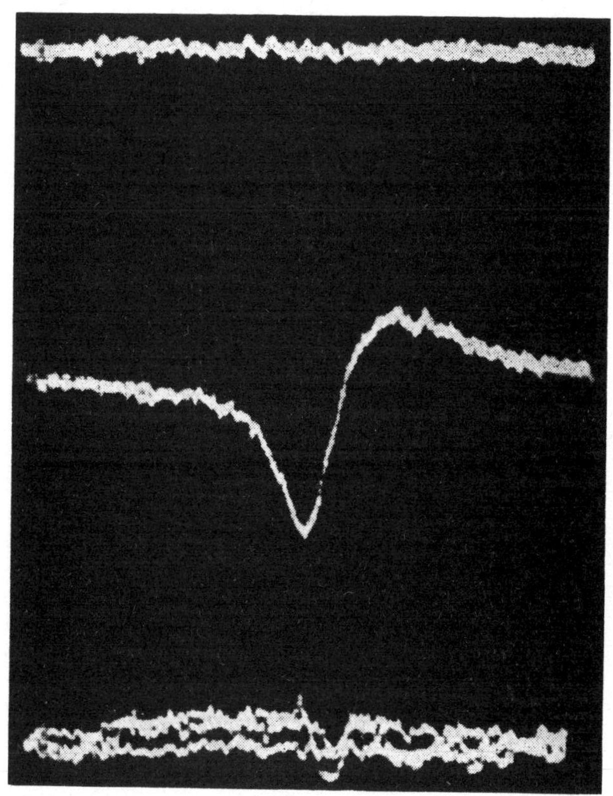

Figure 3.6 Demonstration of the Overhauser effect in ^{7}Li. From Carver and Slichter, *Phys. Rev.* **92**, 213 (1953); © The American Physical Society. The oscilloscope picture shows nuclear absorption plotted vertically as a function of the magnetic field strength on the abscissa. The magnetic field swing is about 0.2 G. The top line shows the normal ^{7}Li nuclear resonance (lost in noise at the 50 kHz frequency of the nuclear magnetic resonance apparatus). The middle trace shows the ^{7}Li nuclear resonance enhanced by saturating the electron spin resonance. The bottom line shows the proton resonance from a glycerin sample containing eight times as many protons under the same experimental conditions used as reference, from which one concludes the ^{7}Li nuclear polarisation was increased by a factor of 100.

mechanism for the nuclei is a kind of cross-relaxation or double-spin-flip process in which the spin vectors of a nucleus and an electron simultaneously flip in opposite directions. The interactions between the electron spin magnetic moment and the nuclear spin magnetic moment is that of hyperfine coupling with an interaction Hamiltonian

$$\mathscr{H}_{\text{int}} = (8\pi/3)\, \beta\beta_{\text{N}}\, gg_{\text{I}}(\mathbf{I}_j \cdot \mathbf{S}_k)\, \delta\, (r_{jk}), \tag{3.37}$$

where β and β_{N} are respectively the electronic and nuclear Bohr magnetons, g and g_{I} the gyromagnetic ratio of the electron and of the nucleus respectively, and r_{jk} is the distance from nucleus j to electron k. The expression (3.37) is of the 'Fermi' or 'contact' type and represents an interaction in which $I_z + S_z$ is conserved.

A transition in which $\Delta S_z = +1$, implies $\Delta I_z = -1$, and vice versa; and this fact is essential to the existence of the Overhauser effect.

In treating the interaction of a conduction electron with a nuclear spin the relevant distribution function can be written as

$$\exp\,[-\bar{k}^2 h^2/8\pi^2 m^* kT_k]\, \exp\,[-g\beta H S_z/kT_{\text{Z}}]\, \exp\,[g_I \beta_N H I_{\text{Z}}/kT_I],$$

where T_{Z} is the electron Zeeman temperature, T_k is the temperature of the electronic translational motion, which may be identified with the room temperature, and T_I is the Zeeman temperature for the nuclei which, taking into account the interaction (3.37), turns out to be given by

$$1/T_I = (g\beta + g_I \beta_{\text{N}})/g_I \beta_{\text{N}} T_k - (g\beta/g_I \beta_{\text{N}} T_{\text{Z}}). \tag{3.38}$$

If the electronic spin is completely saturated (equal population in its upper and lower states) the temperature T_{Z} can be regarded as infinite. Furthermore β is enormously greater than β_{N} so equation (3.38) reduces to

$$T_I = T_k (g_I \beta_{\text{N}}/g\beta).$$

In the case of ^{7}Li, $g\beta$ is $1690\, g_I \beta_{\text{N}}$ and therefore one should have

$$T_I = T_k/1690.$$

The Overhauser effect thus offers the possibility of an enormous reduction in T_I, and hence an enormous enhancement of the nuclear resonance absorption, which is proportional to

$$N_I - N_{I-1} \simeq g_I \beta_{\text{N}} H/[kT_I(2I+1)].$$

The enhancement factor observed by Carver and Slichter was about 140. The reason for this is that the electron spin resonance is not being completely saturated and because nuclei exchange energy with the lattice.

The Overhauser effect is therefore based on the creation, by pumping, of a high temperature in part of the magnetic system, while a good spin-lattice relaxation keeps another part of the system at a low temperature.

The thermodynamic interpretation of the Overhauser effect was first given by Brovetto and Cini[52] and further developed by Barker and Mencher[53]. The effect influenced Bloembergen in his later proposal of the three-level maser (see Chapter 4, Section 5).

In some cases, in fact, the Overhauser effect is very much like a maser action, with the electronic transition as the pump transition, and the nuclear population strongly inverted. This depends on the sign of g_I. In lithium, g_I is positive and the experimental effect was a greatly enhanced but not inverted nuclear population difference. Inversion of an NMR signal via the Overhauser effect was obtained much later in 1960 in silicon[54].

3.8 Spin echo

In the course of nuclear resonance experiments many of the concepts which later became essential for the understanding of masers and lasers were introduced and discussed.

Bloembergen, Purcell and Pound[22] considered in their paper the different kinds of homogeneous and inhomogeneous broadening of the absorption line with the subsequent creation of holes in the inhomogeneous line. The concept of *homogeneous* and *inhomogeneous* lines became very important for the understanding of magnetic resonance, masers and lasers. It was clearly stated later by Portis[55]. As applied to a spin system, *homogeneous* broadening mechanisms broaden the response of each individual spin over the whole linewidth, while an *inhomogeneous* effect spreads the resonance frequencies of different individual spins over some range, widening the overall response of the spin system. The main point is that any excitation applied to one spin in a homogeneous broadened system is immediately transmitted to and shared with all the other spins; in the inhomogeneous case those spins having one particular resonance frequency can be excited without transferring this excitation to other spins having slightly different resonance frequencies under the overall linewidth. Spin-lattice relaxation, dipolar broadening between like spins and exchange interaction are all homogeneous broadening effects. Hyperfine interaction, crystalline defects, and inhomogeneous magnetic fields are all inhomogeneous broadening mechanisms.

In atomic or molecular lines the Doppler effect, due to thermal motion of the particles, produces a homogeneous broadening, while collision effects in a gas give rise to an inhomogeneous broadening. In a real situation, factors producing both homogeneous and inhomogeneous broadening may be present at the same time.

Spin echoes were discussed by E L Hahn, when he was still a graduate student, at a meeting of the American Physical Society in Chicago in November 1949[56].

In nuclear magnetic resonance phenomena any continuous Larmor precession of the spin ensemble which takes place in a static magnetic field is finally interrupted by field perturbations due to neighbours in the lattice. The time for which this precession maintains phase memory has been called the spin–spin, or total relaxation, time t_2. If, at the resonance condition, the ensemble at thermal equilibrium is subjected to an intense RF pulse which is short compared to t_2, the macroscopic magnetic moment due to the ensemble acquires a non-equilibrium orientation after the driving pulse is removed. Bloch[24], on this basis, pointed out that a transient nuclear induction signal should be observed immediately following the pulse as the macroscopic magnetic moment precesses freely in the applied static magnetic field. E L Hahn[57] first verified this behaviour; he was then led to consider a closely related effect which he named *spin echo*[56] which may be described as follows. A resonant oscillatory pulse is applied for a short time and then removed. Signals from the previous free Larmor precession are observed for a short time, but these rapidly disappear in a time of the order of t_2. At a time τ later where $t_2 \ll \tau \ll t_1$, a second resonant radio-frequency pulse is applied and then removed. It is then found that at a time τ after the application of the second pulse a radio-frequency pulse is induced in the receiving circuit. This signal, which apparently arises spontaneously at a time 2τ after the initial pulse is applied, he called a spin echo.

The origin of this spin echo can most easily be understood by considering the special case in which the initially applied pulse is of just sufficient magnitude and duration to redirect the resultant magnetisation through 90° from being parallel to the external magnetic field to being perpendicular to that field. After a further time t_2, however, because of transverse relaxation phenomena or perhaps field inhomogeneities, those nuclear moments which had just been aligned in one direction in a plane perpendicular to the external field will be pointing in all directions in that plane. For intervals of time of short duration compared with t_1, however, the moments will still remain aligned in that plane. If, at time τ after the first pulse, a second pulse is applied whose magnitude (to simplify the discussion) is just double that of the original pulse, then this entire plane will be rotated through 180° (twice the original 90°)[58]. The nuclear moments will therefore tend to unwind their loss of phase exactly (resulting from the transverse relaxation phenomenon) after a time τ, i.e. after the time at which they originally got out of phase.

The method was convincingly compared by N Ramsey[38] with having a number of runners all of whom run at different but constant speeds. If they were started in one direction they would soon be spread out because of their different speeds. However, if, at a time τ after the start, each runner simultaneously reversed his direction, one would find that, at a time after the start, all were neatly drawn up abreast at the starting line.

In figure 3.7 the formation of a spin echo by means of a $\pi/2 - \pi$ pulse sequence is viewed in the rotating reference frame. In figure 3.7(a), before the application of the pulse, the magnetisation M_0 is in thermal equilibrium lying along the

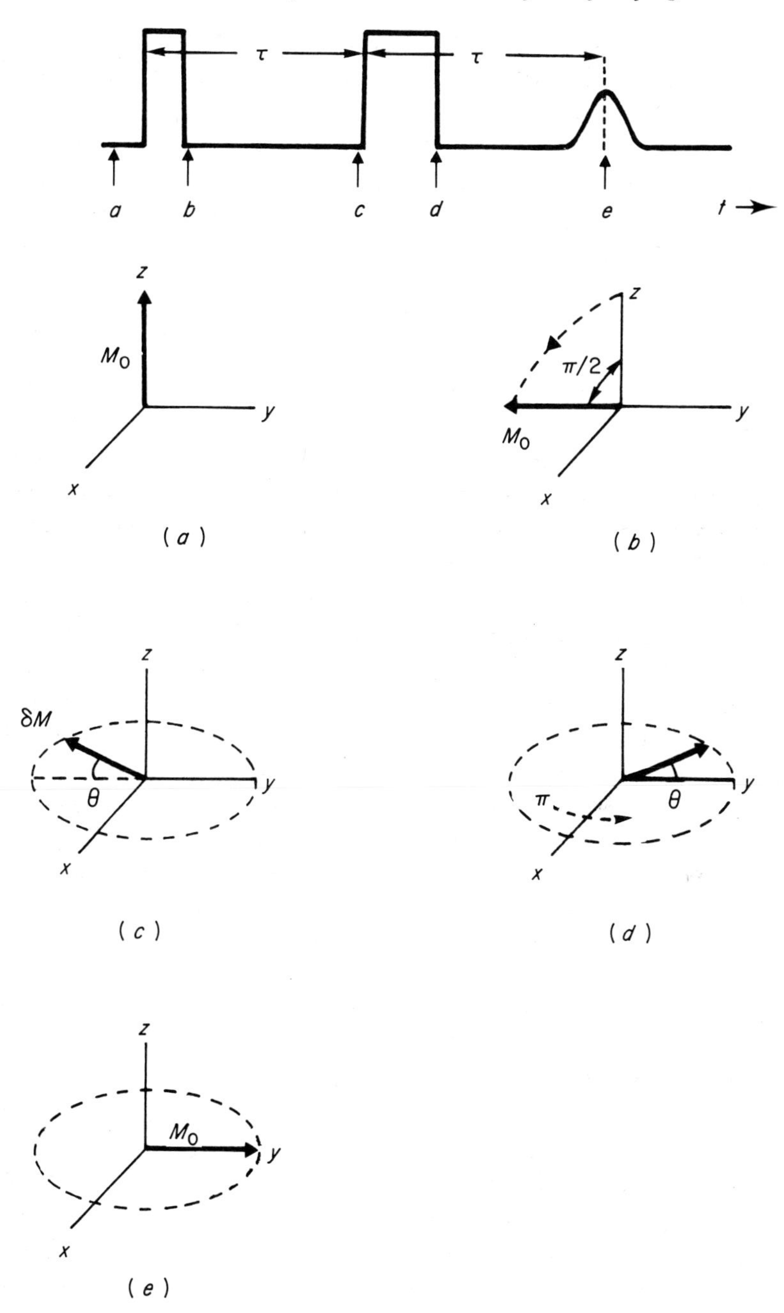

Figure 3.7 The formation of a spin echo by means of a $\pi/2-\pi$ pulse sequence viewed in the rotating reference frame. On top the two pulses and the echo as a function of time. The remaining frames show the position of magnetisation at the different times marked on the upper time scale.

z direction (that is, the direction of the static magnetic field). Immediately after the $\pi/2$ pulse, the magnetisation M_0 has rotated to lie along the positive y axis (figure 3.7(b)). During the time τ the various spins dephase. A group of them, with magnetisation δM is shown in figure 3.7(c), precessed by an extra angle θ. Figure 3.7(d) shows the effect of the π pulse on δM.

Noting the orientation of δM, we see immediately that during a second time interval τ, δM will again advance through the same angle θ, which will bring it exactly along the positive y axis at $t = 2\tau$ (figure 3.7(e)). The argument applies to all spins, because the result does not depend on the angle of advance.

If one takes into account the relaxation times t_1 and t_2 one can see that during the time interval 2τ the magnetisation decays exponentially, so that the size of the magnetisation producing the echo signal is

$$M(t) = M_0 \exp(-2\tau/t_2).$$

Due to the relaxation t_1 after the first pulse, a z component of the magnetisation develops that is subsequently inverted after the second π pulse and does not contribute to the component of M in the x-y plane existing at time τ. If the second pulse is a $\pi/2$ pulse, as in Hahn's original experiment, it can be seen that several echoes are produced.

In liquids, the diffusional motion allows a nucleus to move between different parts of the sample, where the precession rates may differ. As a result, during a spin echo the dephasing during the first interval τ may differ from the rephasing during the second interval τ, and so the echo is diminished. This effect, which is of great practical use as a means of measuring diffusion rates in liquids, was discussed by Hahn in his publication, where he showed that diffusion led to a decay of the echo peak magnetisation M, given by

$$M(t) = M_0 \exp[-\gamma^2(\partial H/\partial z)^2(2D\tau^3/3)],$$

where D is the self-diffusion coefficient of the spin-containing molecule. Hahn discovered spin echoes experimentally, but was soon able to derive their existence from the Bloch equations. This solution, which showed the exponential decay of the echo amplitude with t_2, provided a way of measuring linewidths much narrower than the magnetic inhomogeneity. Understanding the physical basis of echo formation has led to a much deeper insight into resonance phenomena in general and pulse work in particular.

Later, in 1957, Feynman (1918–), Vernon and Hellwarth (1930–)[59] introduced a formalism which establishes a formal similarity between the response of a two-level atomic system subjected to a harmonic electric field and that of a magnetic spin in a combined DC and RF magnetic field.

With their *geometrical representation*, as they called it, it is possible to visualise the atomic dipolar behaviour in terms of the conceptually more simple spin precession.

The Schrödinger equation was written, after a suitable transformation, in the form of the real three-dimensional vector equation

$$\mathrm{d}\mathbf{r}/\mathrm{d}t = \boldsymbol{\omega} \times \mathbf{r},$$

where the components of the vector **r** uniquely determine ψ of a given system and the components of $\boldsymbol{\omega}$ represent the perturbation. In this paper the method was shown to enable the analysis of masers and radiation damping, but it can be applied to the understanding of Dicke superradiance and photon echoes, as well as being a quite general one. Photon echoes in optical transitions were also detected much later[60].

3.9 Optical pumping

Optical pumping is directly connected with the problems we have just been considering. It is a method for producing important changes in the population distribution of atoms and ions between their energy states by optical irradiation. Optical pumping techniques were developed by Kastler, Brossel and others, but it is usually agreed that A Kastler was the leading scientist in the field.

Kastler was born in Guebwiller, Alsace on 3 May 1902. He studied at the École Normale Supérieure, 1921-6, then taught in secondary schools. In 1931 P Daure of Bordeaux University offered him an assistant professorship in his laboratory. There Kastler took his *doctorat des sciences* in 1936. From 1938 to 1941 he was Professor of Physics at Bordeaux University, and he then returned to the École Normale Supérieure in Paris. He was awarded the Nobel Prize for physics in 1966.

In 1949, in collaboration with Brossel, he described an optical method which enabled the redistribution of the populations in the sub-levels of an excited atomic level[61]. The following year he named this method *optical pumping* in a publication in *Journal de Physique et le Radium*[62]. He used this term to describe a process which produced in a stationary form, a situation in which the population of a set of atomic sub-levels (Zeeman levels or hyperfine levels) of the fundamental state was different from the normal Boltzmann distribution. The technique uses a cycle which entails the absorption of optical resonance light followed by its spontaneous re-emission: the basic principle involves the conservation of angular momentum in both the matter and radiation interactions.

The changes in population can be monitored by noting either the change in intensity of the light transmitted by the sample in which optical pumping is produced or the change in either intensity or polarisation of the scattered resonance light. The methods of optical pumping and of optical detection can also be used either together or separately to investigate excited states of atoms. This was accomplished by Brossel and Bitter (1902-1967) in 1952[63] who obtained selective excitation of Zeeman sub-levels of the excited state 6^3P_1 of a mercury atom, and detected the change of polarisation of the re-emitted resonance radiation. This method of studying excited states was called *double resonance* by its authors.

In order to understand the principle of the method proposed by Kastler, let us consider the case of a sodium atom in its ground state, which is $^2S_{1/2}$, with its electron spin $\frac{1}{2}$ split by a magnetic field into two Zeeman sub-levels: $m=-\frac{1}{2}$ and $m=+\frac{1}{2}$. For simplicity let us disregard nuclear spin. By absorption of optical resonance radiation (the D_1 and D_2 lines of sodium), the atom is raised to the $^2P_{1/2}$ and $^2P_{3/2}$ states which are the excited states nearest to the ground state.

Figures 3.8(*a*), (*b*)[64] show the Zeeman structure of the levels involved and the spectral transitions between them. Figure 3.8(*a*) is an energy scheme – the energy of the state is given by the height of the horizontal line representing the state. Spectral transitions are indicated by vertical arrows. Figure 3.8(*b*) is a

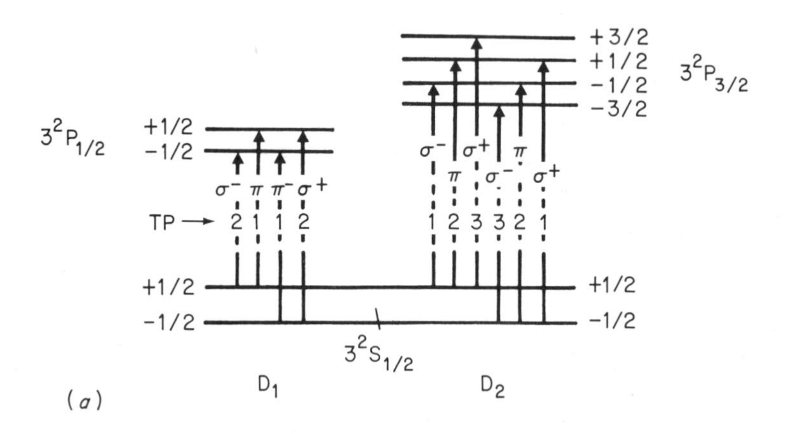

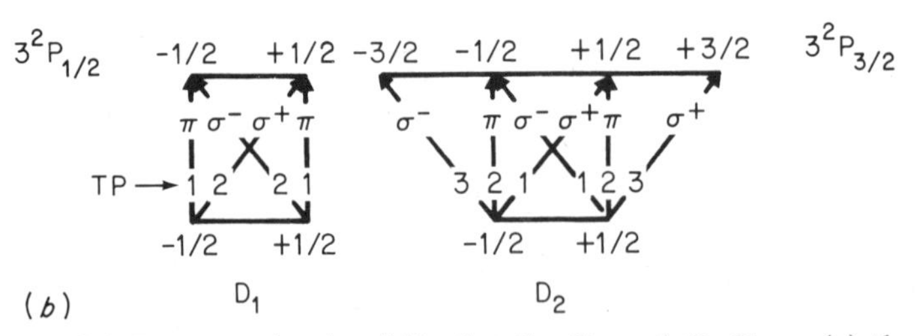

Figure 3.8 Pertinent levels of Na for the D_1 and D_2 lines: (*a*) the energy levels; (*b*) a polarisation diagram. (From C Cohen-Tannoudji and A Kastler, *Progress in Optics* vol 5 edited by E Wolf (North-Holland: Amsterdam, 1966) p3; reprinted by permission.)

polarisation scheme: magnetic sub-levels of the same state are represented by equidistant points on a horizontal line. The arrows indicate the Zeeman transitions.

In this scheme, vertical arrows correspond to $\Delta m = 0$ or π transitions, arrows with a positive slope to $\Delta m = +1$ or σ^+ transitions and arrows with a negative slope to $\Delta m = -1$ or σ^- transitions. The numbers indicated by TP are the relative transition probabilities on an arbitrary scale. Only $\Delta m = 0$ and $\Delta m = \pm 1$ transitions are allowed.

Suppose we illuminate the atoms of an atomic beam of sodium with the circularly polarised yellow light of a sodium lamp. Suppose also that the incident light contains only the D_1 line. Only the Zeeman component σ^+ is exciting the atoms, and in the excited states only $m > 0$ states will be reached. From there the atoms may fall back to the state from which they came. Others will instead make the transition to the lower $m = +\frac{1}{2}$ state.

If this process is repeated several times, atoms of the ground state will leave the $m = -\frac{1}{2}$ level and will accumulate in the $m = +\frac{1}{2}$ level. This change of population can be detected optically.

Take a Na beam (figure 3.9) and orient the atoms in the A region by illumination with σ^+ light. In the B region the atoms are illuminated again with π light, and we measure the ratio of the intensities, $I\sigma^+$ and $I\sigma^-$, emitted. This ratio gives the degree of orientation.

The number of atoms excited during time Δt can then be calculated: it is

$$\Delta n_1' = B_{mm'} \rho_\nu n_m \Delta t,$$

where n_m is the number of atoms in the initial level m, ρ_ν is the spectral density

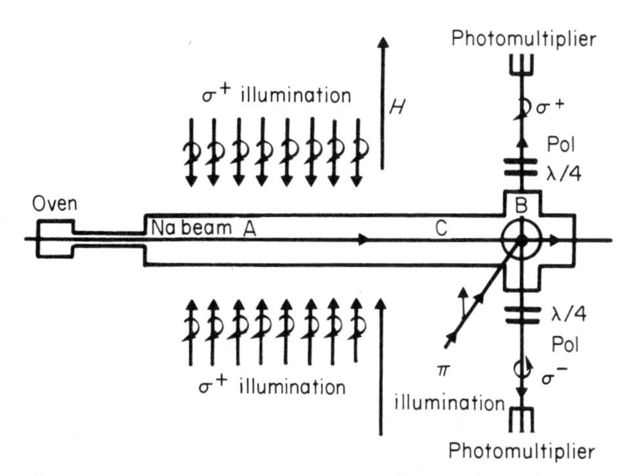

Figure 3.9 Production and detection of atomic orientation in a beam of Na atoms. (From A Kastler, *J. Opt. Soc. Am.* **47**, 460 (1957); reprinted by permission of the Optical Society of America.)

of the incident radiation and $B_{mm'}$ the Einstein absorption probability of the $m \to m'$ Zeeman transition.

The number of atoms which fall back to the fundamental state is

$$-\Delta n_2' = n_m' \Delta t \sum_m A_{m'm} = (n_{m'}'/\tau)\,\Delta t,$$

$A_{m'm}$ being the Einstein emission probability of a transition $m' \to m$ and τ the lifetime of the excited level.

At equilibrium, under the influence of steady radiation, we have

$$\Delta n_1' + \Delta n_2' = 0,$$

from which

$$n_{m'}'/n_m = \rho_\nu \left(B_{mm'} \Big/ \sum_m A_{m'm} \right) = \rho_\nu\,(c^3/8\pi h\nu^3) \left(A_{m'm} \Big/ \sum_m A_{m'm} \right).$$

For the excitation to the sub-level $m' = +\frac{1}{2}$ of the excited state of atoms coming from the sub-level $m = -\frac{1}{2}$ of the fundamental state, the expression $A_{m'm}/\Sigma_m A_{m'm}$ is equal to $\frac{2}{3}$ for D_1 and to $\frac{1}{3}$ for D_2. Therefore the ratio $n_{m'}'/n_m$ of excited atoms depends essentially on ρ_ν, i.e. on the intensity of the effective incident irradiation. By taking the factor $\rho_\nu c^3/8\pi h\nu^3 \simeq 10^{-4}$, the number $\Delta n''$ of atoms which make the transition during the time Δt towards the lower sub-level m'' (different from the initial level m) was calculated by Kastler[62] as

$$\Delta n'' = n_{m'}' \Delta t A_{m'm''} = n_{m'}' (\Delta t/\tau) \left(A_{m'm''} \Big/ \sum_m A_{m'm} \right).$$

$A_{m'm''}/\Sigma_m A_{m'm}$ for the $m' = +\frac{1}{2} \to m'' = +\frac{1}{2}$ transition is equal to $\frac{1}{3}$ for D_1 and to $\frac{2}{3}$ for D_2. The increase in the number of atoms in level $m = +\frac{1}{2}$ of the fundamental state turns out to be $\Delta n'' = \frac{2}{9} \times 10^{-4}\, n_m \Delta t/\tau$, and it is equal to $-\Delta n$, the simultaneous decrease of the number of atoms in the level $m = -\frac{1}{2}$.

By admitting that the two exciting lines D_1 and D_2 have the same intensity, the number of atoms of this level decreases under the influence of irradiation, according to the differential relation

$$\Delta n/n = -\tfrac{4}{9} \times 10^{-4} (\Delta t/\tau).$$

Kastler defined

$$\tau' = \tfrac{9}{4} \times 10^{+4} \tau$$

$$n_{-1/2} = n_0 \exp(-t/\tau') \qquad n_{+1/2} = n_0 [2 - \exp(-t/\tau')].$$

While the number of atoms on level $m = -\frac{1}{2}$ decreases exponentially, the number of atoms of the other level, $m = +\frac{1}{2}$, tends exponentially to twice the initial number. After a time of the order of τ' the ratio $n_{+1/2}/n_{-1/2}$ will have surpassed the value 1. The population asymmetry will be considerable.

The lifetime of level 3^2P of Na is of the order of $\tau = 10^{-8}$ s. Therefore $\tau' = 2 \times 10^{-4}$ s. Kastler considered that, near 100 °C, the thermal velocity of sodium atoms is of the order of 500 m s^{-1}. During the time τ' these atoms travel a distance of some 12 cm. It is therefore possible, by irradiating an atomic beam from the side, to obtain, after a passage of several cm, an important population asymmetry of level m of spatial quantification.

It may be worthwhile at this point to recall briefly the state at this time of research into the detection of radio-frequency resonances by using optical resonance transitions of atoms.

The first researchers to observe the effect of a radio-frequency field on the optical resonance radiation of atoms were E Fermi (1901–1954) and F Rasetti (1901–)[65] and G Breit and A Ellett[66], who in 1925 observed the change of population of resonance fluorescence of mercury vapour by changing the frequency of an applied alternating magnetic field.

If we assume the mercury atom to be a damped classical oscillator, it will have a precession motion around the magnetic field with a Larmor frequency

$$\omega_L = eH/2mc.$$

If the reciprocal of the lifetime of the oscillator (for mercury it is $1/\tau = 10^7$ s^{-1}) is of the same order of magnitude as the Larmor frequency, this precession motion in a static magnetic field makes it possible to obtain emission with the electric field of the wave vibrating in different directions, if excitation of fluorescence is achieved with linearly polarised light. In other words one has depolarisation.

If an alternating field is used at a frequency much larger than the Larmor frequency, ω_L, no large effect on polarisation will occur, the precession being first in one direction and then in the opposite direction, depending on the direction of the alternating field, and also being very small in both directions due to the rapid changing of field. If, on the other hand, the frequency is lower than ω_L, the oscillator will have time to precess in the field before its direction changes and consequently a depolarisation will appear.

In the Fermi and Rasetti experiments[65] the magnetic field could be changed from 1.13 to 2.13 G (1 G gives a Larmor precession frequency of about 1.4×10^6 s^{-1} for a classical oscillator). The frequency of the field could be changed from 1.2 to 5×10^6 s^{-1}. At 1.13 G and a frequency of 5×10^6 s^{-1} they found that practically no depolarisation was present, whilst at 1.87 G at the same frequency a depolarisation was observed. At 2.13 G the depolarisation was as large as in a static field. If the field strength had been larger by a factor $\frac{3}{2}$, the results obtained could have been explained satisfactorily on the basis of the classical oscillator model.

They therefore measured the *g* factor and the lifetime of the excited states. These results were described in a short letter of 17 lines to *Nature*[65]. A more detailed description of the experiment was published later[67].

Rasetti remembers the experiment as follows[68]:

> Wood and Ellett† and Hanle‡, had announced their remarkable discovery of the effects of weak magnetic fields on the polarisation of the resonance radiation of mercury. Rasetti had observed these effects in Florence. When Fermi came to that university a few months later, he was greatly interested in the phenomenon, whose only theory at the time was a classical one based on the concept of Larmor precession. Fermi pointed out that, since the mercury resonance line showed an anomalous Zeeman effect with a Landé factor of $\frac{3}{2}$, the mercury atom should more likely precess with a frequency $\frac{3}{2}$ times higher than the Larmor frequency. The choice between the two alternatives might be decided by investigating the behaviour of the polarisation under magnetic fields, of the intensity of about one gauss and a frequency of a few megacycles per second in approximate resonance with the precession frequency of the atom.
>
> Rasetti had experience with the spectroscopic technique, but neither of the experimenters had any with radio-frequency circuits. However, Fermi calculated the characteristics of a simple oscillator circuit which should produce fields of the proper strengths and frequencies. Fortunately some triodes were discovered in an instrument cabinet and pronounced by Fermi apt to operate the projected circuit. The laboratory also possessed several hot-wire ammeters to measure the current in the coils, in order that the magnetic field strength could be determined. Had these instruments not been available, the experiment could not have been performed, since the research budget of the laboratory was exceedingly meagre and did not allow the purchase of costly equipment. Another consequence of this financial situation was the fact that the building was never heated, since it was easy to calculate that one month's heating would have absorbed the entire annual budget of the Physics Department. The temperature in the building from December to March ranged from 3 to 6 °C. Unfortunately in the spring, when the experiments were performed, the room temperature had risen to 12 °C, more comfortable for the inhabitants, but somewhat too high to ensure a sufficiently low density of the saturated mercury vapour.
>
> Inductance coils and other simple parts were built by the experimenters and when the circuit was assembled, it instantly worked as Fermi had

† A Ellett, *Nature* **114**, 931 (1924).
‡ W Hanle, *Z. Phys.* **30**, 93 (1924).

predicted. The experiments were readily performed; unfortunately the accuracy was poor, due to the high temperature and the photographic method employed for measuring the polarisation. Still, the results clearly showed that the precession frequency of the atom agreed with the prediction based on the Landé factor.

In the following years much experimental work was done on atomic fluorescence (for example see notes 69, 70). Then microwave and radio-frequency spectroscopy started to catch the attention of physicists and chemists. The magnetic resonance which had been detected in atomic and molecular beams was now detected in bulk materials; the molecular rotational transitions and the nuclear electric quadrupole interactions began to be observed in the microwave spectra of the vapours of various compounds. These new experiments produced many advances in the elucidation of atomic and molecular structure. These works permitted the determination of the nuclear spins and moments which, at that time, were two of the few quantities which could be obtained experimentally and so could be used in any discussion on nuclear structure.

In 1949 Bitter (1902–1967) called attention to the possibility of optical detection of a resonance at radio-frequency[71]. However, Pryce[72] showed that the detection method proposed by Bitter was not possible. It was at this time that Brossel and Kastler[61, 62] proposed a method for the optical re-orientation of atoms and discussed the different ways of detecting orientation. Jean Brossel recalled those days:[73]

> It was the end of 1948. I had already been in England for three years, at the Tolansky laboratory in Manchester, where I had become acquainted with interferometry problems and with the optical methods of studying hyperfine structures. One of our American colleagues told Kastler of the existence of a bursary for a young physicist who might be interested in any aspect of spectroscopy....I had to make up my mind very quickly and we agreed I would go to work at Francis Bitter's laboratory.

The work Brossel was called upon to do was on an experiment which could allow the extension of the magnetic resonance methods to the study of excited atomic states. Brossel tells us that after hesitations and changes...

> I prepared the experiment. A few days before it was ready, M H L Pryce (to avoid a misadventure to some experimentalist) published a paper in *Physical Review* where he gave the right answer to the problem posed by Bitter,

and where he found that the experiment proposed by Bitter was not valid.

It was no longer worthwhile to proceed in this direction; everything had to begin again. It was too late to go back on my decision to stay for one more year at MIT. I had to make the best use of my time.

I began to read texts again, particularly the books by Pringsheim† and Mitchell and Zemansky‡ on fluorescence and optical resonance. And when I began to read afresh the chapter on the polarisation of optical resonance light, I understood that there was a very simple and powerful method for the observation of the magnetic resonance of excited levels; it was only necessary to prepare the system by exciting it with *polarised* light. A situation is then obtained in which large differences exist in the populations of the Zeeman sub-levels of the excited state.

At resonance, $\omega = \omega_0$, the radio-frequency field re-orients the kinetic moment and equalises the populations. This results in a depolarisation of fluorescence light; no optical resolution is needed. I immediately told Francis of this. We decided to perform the experiment immediately.

During this period I was in constant contact with Kastler in Paris, keeping him in touch, through regular correspondence, with our progress and work. As often is the case where the situation is ripe for discovery, our ideas on the solution of the problem had evolved in parallel. By coincidence we had arrived independently and at the same time at an identical conclusion. A week after I had discussed it with Francis, I received a letter from Paris in which Kastler was proposing exactly the same experiment.

He was developing his idea in connection with the sodium atom, employing an excitation with *circularly* polarised light (which led him a few months later to propose the idea of optical pumping) whilst I had 'seen' the effect on the 2537 line of the even isotope of Hg with π excitation.

As we were sure of the success of the experiment we published the results immediately in *Comptes Rendus de l'Academie des Sciences, Paris*, not omitting to point out that excitation by a directed beam of slow electrons would allow the extension of the method to *all* excited atomic states.

In a few months the experiment performed at MIT with Paul Sagalyn completely confirmed the preceding conclusions. The lack of space, which is a factor in every laboratory, had forced us to operate in highly uncomfortable conditions, being only too happy to be able to make use of a kind of loft without windows that was used as a box-room and that had long been abandoned, under the exit stairway of the Eastman Building, towards Walker's Hall...During the time all this work was

† See note 70.
‡ See note 69.

> being undertaken at MIT, Kastler was trying to extend the optical detection methods of magnetic resonance to ground states and invented optical pumping... Kastler quickly started to assemble an experimental set-up to put into practice, on an atomic jet, the production of an atomic orientation of the ground state of Na by optical pumping. It was a failure. The same happened to me when, a few months before I left MIT, I tried to produce nuclear orientation of ^{199}Hg by pumping the vapour using the 2537 line. The intensity of my sources was too low. A few years later in Paris, when I built up some very bright sources, I did this experiment again with Cagnac, and we succeeded without difficulty.

The experiments described by Brossel were performed with the $3P_1$ state of the isotopes of mercury by Brossel and Bitter[63] and on the fundamental state of ^{23}Na by Brossel, Kastler and Winter[74] and Hawkins and Dicke[75]. The experiment with Sagalyn was published in 1950[76], and the one with Cagnac in 1959[77].

It is also worth noting that by 1936 in his doctoral thesis on grating (echelon) excitation of mercury vapour, Kastler[78] in showing that selective excitation of Zeeman sub-levels of excited states could be obtained by a suitable polarisation of monochromatic exciting radiations, was already proposing and realising experiments whose basic ideas were not very different from the ones that conducted him, 14 years later, to the discovery of optical pumping.

Apparently, however, Kastler had never had it in mind to use optical pumping for population inversion. In the summary of his paper[62] he writes:

> By illuminating the atoms of a gas or of an atomic beam with oriented resonance radiations (light beams having a determined direction) which are suitably polarised, it is possible, when these atoms in the ground state are paramagnetic (quantum numbers $J \neq 0$ or $F \neq 0$) to obtain an unequal population of the different m sub-levels that characterise the spatial or magnetic quantification of the ground level. A rough evaluation shows that, with present irradiation facilities, this population asymmetry can become very important. From the examination of the transition probabilities of Zeeman transitions π and σ it can be seen that illumination with natural or linearly polarised light enables the concentration of atoms either in the m sub-levels of the middle ($m = 0$) or, on the other hand, on the external sub-levels (m maximum).
>
> The use of circularly polarised light enables the creation of a population asymmetry between negative and positive m levels, with the sign of this asymmetry able to be reversed by reversing the direction of the circular polarisation of the incident light.
>
> This creation of asymmetry can be obtained either in the absence of any external field or in the presence of a magnetic or electric field. In the presence of an external field the different sub-levels m (in the case of a

magnetic field) or $|m|$ (in the case of an electric field) are energetically different, and the creation of a population asymmetry by the optical process corresponds to an increase or a decrease of the 'spin temperature'.

A population asymmetry of the sub-levels m of the ground state can be detected optically by examination of the intensity of polarisation of optical resonance radiations. The use of photoelectric detectors and a modulation technique enables easy and sensitive detection.

The optical examination of the different branches into which an atomic beam divides itself in the Stern and Gerlach experiment permits the control of quantum level m of the atoms of each of these branches. This optical method allows extension of the magnetic analysis of atoms in the Stern and Gerlach experiment to the study of metastable excited levels.

In the magnetic resonance experiments the transitions induced by the radio-frequency oscillating magnetic field tend to destroy the population inequality of levels m. The study of the magnetic resonance of the atoms of an atomic beam can therefore be undertaken by replacing the non-uniform magnetic fields of the Rabi set-up by an optical producer of an asymmetry which precedes the magnetic resonance set-up, and by an optical detector with an asymmetry at the output of the resonator. The optical method permits extension of the study of magnetic resonance to metastable levels.

This method enables one to study the transitions between hyperfine levels in zero field, the hyperfine Zeeman effects in the case of low fields and the hyperfine Paschen–Back effects in strong fields. So, thanks to the connection between the hyperfine Zeeman effect and the hyperfine Paschen–Back effect, one can optically analyse pure nuclear resonance in fields that decouple **J** and **I** vectors. Finally, the study of the Stark effect of an atomic level by the resonance method can equally well be undertaken optically. The process for the optical study of an atomic beam enables large beams with ill defined boundaries to be used. The set-up to be necessary for this study is therefore simple and cheap . . .

At this point the various consequences of the method proposed by Kastler with respect to the problems involved in laser developments become apparent. Kastler proposed a method of a magnetic type to change the population of some levels, with the intention of using the changes in the characteristics of the emitted light in the transitions among such levels to study magnetic resonance phenomena. He never did mention the possibility of obtaining population inversion in this way, nor the possibility of using this inversion to create amplifiers of light generation. Neither did he give any priority to the invention of the laser, although today it is in fact by the use of optical pumping methods that some lasers work.

Notes

1. A Kastler, *Cette étrange matière* (Stock: Paris, 1976)
2. The explanation of magnetism in classical terms may be attributed to Weber (*Leipzig Berichte*, I, 346 (1847); *Ann. Phys., Lpz.* LXXIII, 241 (1848) [translated in Taylor's *Scientific Memoirs* V, p. 477]; *Abh. K. Sächs. Ges.* I, 483 (1852); *Ann. Phys., Lpz.* LXXXVII, 145 (1852) [translated in Tyndall and Francis' *Scientific Memoirs*, p. 163]; *Leipzig Abh. Math. Phys.* X, 1 (1873); *Phil. Mag.* XLIII, 1 and 119 (1872). The re-statement of these ideas in terms of the theory of electrons was undertaken in 1901–3 by W Voigt (*Gött. Mach.* 169 (1901); *Ann. Phys., Lpz.* **9**, 115 (1902) and J J Thomson (*Phil. Mag.* **6**, 673 (1903)). The general theory of magnetism received a complete formulation by P Langevin (*Ann. Chem. Phys.* V, 70 (1905) and *C.R. Acad. Sci., Paris* **139**, 1204 (1904)). In 1907 Langevin's theory was extended by P Weiss so as to give an account of ferromagnetism (*Bull. Séances Soc. Fr. Phys.* 95 (1907); *J. Physique* **6**, 661 (1907)). For a quantum theory it was necessary to wait for the development of a concept of spin (W Pauli, *Z. Phys.* **21**, 615 (1920)). The need for a quantum theory had already been demonstrated by N Bohr (N Bohr, *Studieren over Metallernes Elektronteori*, Copenhagen, 1911) and H J Van Leuven (*Inaugural Dissertation*, Leiden, 1919 and *J. Phys. Radium* **2**, 361 (1921)) (see also J H van Vleck in *The Theory of Electric and Magnetic Susceptibilities*, Oxford University Press: Oxford, 1932). The correct expression for the diamagnetism was first obtained by Landau (*Z. Phys.* **64**, 629 (1930)). The concepts of ferromagnetism and antiferromagnetism were developed thanks to the exchange field ideas of P A M Dirac, *Proc. R. Soc.* A **112**, 661 (1926); L Néel (*Ann. Phys., Lpz.* **17**, 5 (1932); *J. Physique* **3**, 160 (1932); *Ann. Phys., Lpz.* **5** 232 (1936); *C.R. Acad. Sci., Paris* **203**, 304 (1936)), W L Heisenberg (*Z. Phys.* **49**, 619 (1928)) and J. Frenkel (*Z. Phys.* **49**, 31 (1928))
3. W Lenz, *Z. Phys.* **21**, 613 (1920)
4. P Ehrenfest, *Comm. Leiden Suppl.* no. 44b (1920)
5. G Breit and H Kamerlingh Onnes, *Comm. Leiden* no. 168c (1924)
6. For more information on this point see J C Verstelle and D A Curtis, *Paramagnetic Relaxation* in *Handbuch der Physik* edited by S Flugge (Springer Verlag: Berlin, 1968) vol. 18/1, p1
7. W Majorana, *Nuovo Cim.* **9**, 43 (1932)
8. I I Rabi, *Phys. Rev.* **51**, 652 (1937)
9. I I Rabi, S. Millman, P. Kusch and J. R. Zacharias, *Phys. Rev.* **53**, 318 (1938); *Phys. Rev.* **55**, 526 (1939)
10. A historical survey together with an exhaustive discussion of these experiments may be found in H Kopferman, *Nuclear Moments* (Academic: New York, 1958)

11 L W Alvarez and F Bloch, *Phys. Rev.* **57**, 111 (1940)

12 I Waller, *Z. Phys.* **79**, 370 (1932)

13 H B G Casimir and F K Dupré, *Physica* **5**, 507 (1937). The theory of paramagnetic relaxation times was mainly developed by Waller (note 12), R Kronig, *Physica* **6**, 33 (1939), J H van Vleck, *Phys. Rev.* **57**, 426 (1940). For a discussion of nuclear paramagnetism see E Teller and W Heitler, *Proc. R. Soc.* **155**, 629 (1936)

14 The concept of magnetic temperature has been discussed in a number of papers. See for example the excellent paper by J H van Vleck, *Nuovo Cim. Suppl.* **6**, 1081 (1957) and Bloembergen's review in *Am. J. Phys.* **41**, 325 (1973) where other useful references may be found. Spin temperature was extensively discussed by N Bloembergen, *Physica* **15**, 386 (1949)

15 C J Gorter and L J F Broer, *Physica* **9**, 591 (1942)

16 C J Gorter, *Physica* **3**, 995 (1936)

17 C J Gorter, *Paramagnetic Relaxation* (Elsevier: Amsterdam, 1947)

18 F Bloch, W W Hansen and M Packard, *Phys. Rev.* **70**, 127 (1946); F Bloch, *Phys. Rev.* **70**, 460 (1946); F Bloch, W W Hansen and M Packard, *Phys. Rev.* **70**, 474 (1946)

19 E M Purcell, H C Torrey and R V Pound, *Phys. Rev.* **70**, 37 (1946)

20 E Zavoisky, *J. Phys. USSR* **9**, 211 (1945); **10**, 197 (1946)

21 Bloch's work on magnetism is well known. Fundamental works are the classic paper on the spin-wave approach and the celebrated $T^{3/2}$ relation for the decrease of saturation magnetisation (*Z. Phys.* **61**, 206 (1930)) and the wall concept in a ferromagnet (*Z. Phys.* **74**, 295 (1932))

22 N Bloembergen, E M Purcell and R V Pound, *Phys. Rev.* **73**, 679 (1948)

23 F Bloch, W W Hansen and M Packard, *Phys. Rev.* **70**, 127 (1946)

24 F Bloch, *Phys. Rev.* **70**, 460 (1946)

25 Equation (3.2) is exactly the Planck condition for the transition between two adjacent Zeeman levels. For a single nucleus of magnetic moment μ and spin I, there are $2I + 1$ levels separated by an energy

$$\Delta E = \mu H/I,$$

so that

$$\Delta E/\hbar = \omega = \mu H/I\hbar = \gamma H,$$

where $\gamma = \mu/I\hbar$

26 A justification of this exponential law was given shortly afterwards by N Bloembergen (note 22)

27 The possibility of representing the oscillating field as the superposition of two fields rotating in opposite directions was pointed out by F Bloch and A Siegert, *Phys. Rev.* **57**, 522 (1940)

28 The use of rotating coordinates for the treatment of magnetic resonance was dealt with by I I Rabi, N F Ramsey and J Schwinger, *Rev. Mod. Phys.* **26**, 167 (1954). Here the complete quantum mechanical treatment of the theorem can also be found.

29 From figures 1 and 2 of note 24

30 F Bloch, W W Hansen and M Packard, *Phys. Rev.* **70**, 474 (1946)

31 W E Lamb and R C Retherford, *Phys. Rev.* **79**, 549 (1950)
32 R V Pound, *Phys. Rev.* **79**, 685 (1950)
33 E M Purcell and R V Pound, *Phys. Rev.* **81**, 279 (1951)
34 See: note 17; L J F Broer, *Physica* **10**, 801 (1943); N F Ramsey, *Nuclear Moments* (John Wiley: New York, 1953); N Bloembergen, *Physica* **15**, 386 (1949)
35 A Abragam and W G Proctor, *Phys. Rev.* **109**, 1441 (1958)
36 N Bloembergen, *Am. J. Phys.* **41**, 325 (1973)
37 J H van Vleck, *Nuovo Cim. Suppl.* **6**, 1081 (1957); B D Coleman and W Noll, *Phys. Rev.* **115**, 262 (1959); P T Landsberg, *Phys. Rev.* **115**, 518 (1959); L C Hebel jr, *Solid State Physics* (Academic: New York, 1963) vol. 15
38 N F Ramsey, *Phys. Rev.* **103**, 20 (1956)
39 N F Ramsey, *Ann. Rev. Nucl. Sci.* **1**, 99 (1952); reproduced with permission © by Annual Review Inc.
40 In a footnote on p. 690 of the paper by J Weber, *Rev. Mod. Phys.* **31**, 681 (1959)
41 J Combrisson, A Honig and C H Townes, *C.R. Acad. Sci., Paris* **242**, 2451 (1956)
42 G Feher, J P Gordon, E Buehler, E A Gere and C D Thurmond, *Phys. Rev.* **109**, 221 (1958)
43 A Overhauser, *Phys. Rev.* **89**, 689 (1953)
44 T W Griswold, A F Kip and C Kittel, *Phys. Rev.* **88**, 951 (1952)
45 A W Overhauser, *Phys. Rev.* **91**, 476 (1953)
46 I am indebted for this information on the Urbana work to Professor C P Slichter who kindly provided me with this recollection.
47 A W Overhauser, *Phys. Rev.* **92**, 411 (1953)
48 See F Bloch, *Phys. Rev.* **93**, 944 (A) (1954); C Kittel, *Phys. Rev.* **95**, 589 (1954); J Korringa, *Phys. Rev.* **94**, 1388 (1954); A Abragam, *Phys. Rev.* **98**, 1729 (1955)
49 See J I Kaplan, *Phys. Rev.* **99**, 1322 (1955)
50 T R Carver and C P Slichter, *Phys. Rev.* **92**, 212 (1953). Figure 1 from this reference is our figure 3.6
51 T R Carver and C P Slichter, *Phys. Rev.* **102**, 975 (1956)
52 P Brovetto and M Cini, *Nuovo Cim.* **11**, 618 (1954)
53 W W Barker and A Mencher, *Phys. Rev.* **102**, 1023 (1956)
54 J Combrisson in *Quantum Electronics*, edited by C H Townes (Columbia University Press: New York, 1960) p167
55 A M Portis, *Phys. Rev.* **91**, 1071 (1953)
56 E L Hahn, *Phys. Rev.* **72**, 746 (1950). A full account of the theory and experiments was given later in *Phys. Rev.* **80**, 580 (1950)
57 E L Hahn, *Phys. Rev.* **77**, 297 (1950)
58 In Hahn's original experiment two equal pulses were applied (or two pulses at 90°). The application of 90° and 180° pulses was first considered by H Y Carr and E M Purcell, *Phys. Rev.* **94**, 630 (1954)
59 R P Feynman, F L Vernon jr and R W Hellwarth, *J. Appl. Phys.* **28**, 49 (1957)
60 N A Kurnit, I D Abella and S R Hartmann, *Phys. Rev. Lett.* **13**, 567 (1964)

61 J Brossel and A Kastler *C.R. Acad. Sci., Paris* **229**, 1213 (1949)

62 A Kastler, *J. Physique Radium* **11**, 255 (1950)

63 J Brossel and F Bitter, *Phys. Rev.* **86**, 308 (1952). An excellent review of these experiments was made later by F Bitter, *Appl. Opt.* **1**, 1 (1962)

64 We closely follow here the description given by A Kastler in *J. Opt. Soc. Am.* **47**, 460 (1957)

65 E Fermi and F Rasetti, *Nature* **115**, 764 (1925); *Z. Phys.* **33**, 246 (1925)

66 G Breit and A Ellett, *Phys. Rev.* **25**, 888 (1925)

67 E Fermi and F Rasetti, *Rend. Lincei* **1**, 716 (1925); **2**, 117 (1925). The work described in the second paper was performed at the Istituto Fisico, Florence

68 F Rasetti in the comment on Fermi's works in *Note e Memorie di E Fermi*, vol. I (Accademia dei Lincei: Rome, 1962) p159

69 A C G Mitchell and M W Zemansky, *Resonance Radiation and Excited Atoms* (Cambridge University Press: Cambridge, 1934)

70 P Pringsheim, *Fluorescence and Phosphorescence* (Interscience: New York, 1949)

71 F Bitter, *Phys. Rev.* **76**, 833 (1949)

72 M H L Pryce, *Phys. Rev.* **77**, 136 (1950)

73 J Brossel, *Quelques souvenirs...* in *Polarisation, Matière et Rayonnement, Volume Jubilaire en l'Honneur d'A Kastler*, (Presses Universitaires de France: Paris, 1969) p143

74 J Brossel, A Kastler and J Winter, *J. Physique Radium* **13**, 668 (1952)

75 W B Hawkins and R H Dicke, *Phys. Rev.* **91**, 1008 (1953)

76 J Brossel, P Sagalyn and F Bitter, *Phys. Rev.* **79**, 196, 225 (1950)

77 B Cagnac, J Brossel and A Kastler, *C.R. Acad. Sci. Paris* **246**, 1027 (1958); B Cagnac, *J. Phys. Radium* **19**, 863 (1958); B Cagnac and J Brossel, *C.R. Acad. Sci. Paris* **249**, 77, 253 (1959)

78 A Kastler, *Thèse*, Paris (1936) and *Ann. Phys., Lpz.* **6**, 663 (1936)

4

The Maser

4.1 Introduction

Population inversions for use in working devices were first obtained in the microwave region, where the spontaneous emission probability, which is proportional to the cube of the frequency, is so small as to be negligible.

C H Townes[1] has always connected these results with the development, during the Second World War, of microwave technology. After the war, a good deal of attention was given to the interaction between microwaves and matter, especially gases. This led to the growth of a new field of research known as *microwave spectroscopy*, which was developed initially in several places (firstly in the industrial laboratories possessing radar apparatus) and then spread quickly to universities.

The maser idea was therefore born as a logical consequence of the resultant detailed knowledge of the interaction between microwaves and matter. In Townes' words[1], 'It was the mixture of electronics and molecular spectroscopy which set appropriate conditions for invention of the maser'. The physical principles and experimental techniques for its development were thus well established in the period 1945–50.

The idea originated independently in the USA in the Universities of Maryland and of Columbia and in the Soviet Union at Moscow's Lebedev Institute, during the early Fifties. The importance of work carried out at Columbia and at the Lebedev Institute, in the field of both masers and lasers, was recognised by the international scientific community by the award of the Nobel Prize for Physics to C H Townes, N G Basov and A M Prokhorov in 1964.

As so often happens, after the report of the first maser operation, other concomitant or prior works were found, which we shall consider later.

4.2 Weber's maser

The first public description of the maser principle (without a working device) was at an Electron Tube Research Conference in Ottawa, Canada, in 1952, by Joseph Weber (1919-)[2]. Weber was then a young electrical engineering professor at the University of Maryland and a consultant at the United States Naval Ordinance Laboratory. He graduated in Annapolis and was first a naval officer. He then went to the Catholic University in Washington DC where he obtained his PhD in 1951. The idea of the maser came to him after he had attended a seminar on stimulated emission by G Herzberg while he was studying for his doctorate at Washington.

In his work, Weber considered a system with two energy levels, E_1 and E_2 ($E_2 > E_1$) with populations n_1 and n_2 respectively[3]. By irradiating with radiation of frequency

$$\nu = (E_2 - E_1)/h, \tag{4.1}$$

the absorbed power in the transitions $1 \rightarrow 2$ is

$$P_{\text{abs}} = W_{12} h\nu n_1. \tag{4.2}$$

The power emitted by particles which decay from the upper state down to the lower state is

$$P_{\text{em}} = W_{21} h\nu n_2. \tag{4.3}$$

Because $W_{12} = W_{21}$ the net absorbed power is

$$P_{\text{net}} = W_{12} h\nu (n_1 - n_2). \tag{4.4}$$

At equilibrium the number of particles in state 2 is governed by the Maxwell-Boltzmann law

$$n_2 = n_1 \exp(-h\nu/kT) \simeq n_1(1 - h\nu/kT). \tag{4.5}$$

By substituting equation (4.5) into equation (4.4) we obtain

$$P_{\text{net}} = W_{12} (h\nu)^2 n_1/kT. \tag{4.6}$$

Weber writes[2]

> ...and this [P_{net}] is a positive quantity. Thus under ordinary circumstances we get absorption of radiation (ordinary microwave spectroscopy) because the transition probability up is the same as the transition probability down, but since there are more oscillators in the lower state, we get a net absorption... We could get amplification if somehow the number of oscillators in the upper states could be made greater than the number in the lower states. A method of doing this is suggested by Purcell's†[3] negative temperature experiment.

He then considered two ways of obtaining the necessary reversal; a sudden reversal of magnetic field or a pulsed system with polar molecules. By applying an electric field one obtains a separation of levels, by the Stark effect. If equilibrium is reached and the field is quickly inverted, one obtains $n_2 > n_1$ for a time corresponding to the thermal relaxation time. Weber also suggested making a gas flow through a region of electric field reversal, to have continuous operation.

The work described above was carried out by Weber in 1951 and presented at conference in 1952. He published it in summary form in 1953[2]. As explained by Weber himself[4] it was his intention to publish his results in a

> ...widely read journal. Early in 1953 Professor H J Reich of Yale University wrote to say that he had been chairman of the 1952 Electron Tube Conference program committee, and was also editor of a (not so widely read) journal. As a result the conference summary report was published in the June 1953 issue of *Transaction of the Institute of Radio Engineers Professional Group on Electron Devices.*

In this work Weber underlines the fact that the amplification is coherent. The method he proposed for obtaining population inversion has never, in fact, been put into practice and it seems most unlikely ever to be so.

Moreover, one may observe that although he uses the word 'coherent' in his work, only an amplifier is considered, and nothing is said on self-sustained oscillation. However, the basic maser ideas – using stimulated emission to excite atoms or molecules and invert the population – are clearly stated.

After his presentation of this work at the conference, Weber was asked by RCA to give a seminar on his idea. For this he received a fee of $50. After the seminar Townes wrote to him, asking for a reprint of the paper. Weber's work was, however, not quoted in the first of Townes' papers but was referred to later[5].

4.3 Townes and the first ammonia maser

The first, experimental, operating maser was built by a group of researchers at Columbia University, headed by C H Townes.

Charles H Townes was born in 1915 in Greenville, South Carolina. When he was 16 he entered Furman University. Although he soon discovered his vocation for physics, he also studied Greek, Latin, Anglo-Saxon, French and German, and received a BA degree in modern languages after three years at Furman. At

† E M Purcell and R V Pound, *Phys. Rev.* **81**, 279 (1951).

the end of his fourth year he received a BSc in physics. He next went to Duke University on a scholarship. When he was 21 he finished work on his master's degree, continuing to study French, Russian and Italian. He then went to the California Institute of Technology in Pasadena, where, in 1939, he received his PhD, after which he accepted an appointment at the Bell Telephone Laboratories. During the war Townes was assigned to work with Dean Wooldridge, who was then designing radar bombing systems. Although Townes preferred theoretical physics, he nevertheless worked on this practical project.

At that time people were trying to push the operational frequency of radar higher. The Air Force asked Bell to work on radar at 24 000 MHz. Such a radar would exploit an almost unexplored frequency range and would result in more precise bombing equipment.

Townes, however, observed that radiation of that frequency is strongly absorbed by water vapour. The Air Force nevertheless insisted on trying it. So the radar was built by Townes, who then was able to verify that it did not work. As a result of this work, Townes became interested in microwave spectroscopy.

In 1947 Townes accepted an invitation from Isidor I Rabi to leave Bell Laboratories and join the faculty at Columbia University in which Rabi was working. There was a Radiation Laboratory group in the Physics Department which had continued the war-time programme on magnetrons for the generation of millimetre waves. This laboratory was supported by a Joint Services contract from the US Army, Navy and Air Force, with the general purpose of exploring the microwave region and extending it to shorter wavelengths. Among the people active in the sponsorship of this programme were Dr Harold Zahl of the Army Signal Corps and Paul S Johnson of the Naval Office of Research. Townes quickly became an authority on microwave spectroscopy and on the use of microwaves for the study of matter[6].

In 1950 he became a full professor of physics. In the same year Johnson organised a study commission on millimetre waves and asked Townes to take the chair. Townes worked on the committee for nearly two years and became rather dissatisfied with its progress. Then, one day in the spring of 1951, when he was in Washington DC to attend a meeting of the committee, he tells us[7]

> By coincidence, I was in a hotel room with my friend and colleague Arthur L Schawlow, later to be involved with the laser. I awoke early in the morning and, in order not to disturb him, went out and sat on a nearby park bench to puzzle over what was the essential reason we had failed [in producing a millimetre wave generator]. It was clear that what was needed was a way of making a very small, precise resonator and having in it some form of energy which could be coupled to an electromagnetic field. But that was a description of a molecule, and the technical difficulty for man to make such small resonators and provide energy meant that

> any real hope had to be based on finding a way of using molecules! Perhaps it was the fresh morning air that made me suddenly see that this was possible: in a few minutes I sketched out and calculated requirements for a molecular-beam system to separate high-energy molecules from lower ones and send them through a cavity which would contain the electromagnetic radiation to stimulate further emission from the molecules, thus providing feedback and continuous oscillation.

He did not say anything at the meeting, and at once went back to Columbia and started work on his new idea. He gathered around him a group of researchers including H J Zeiger, a post-doctoral fellow and James P Gordon, then working for his doctorate.

The active material envisaged by Townes was ammonia gas. In the classical picture, the ammonia molecule (NH_3) is like a triangular pyramid (figure 4.1) with the three hydrogen atoms at the vertices of the base and the nitrogen atom at the apex. A quantum mechanical treatment shows that there are two possible, equivalent positions of the nitrogen atom, either above or below the plane formed by the three hydrogen atoms, and that the potential energy of the nitrogen atom as a function of its distance from this plane is as shown qualitatively in figure 4.2. Here the broken lines show what would be valid if only one side (or the other) of the plane were accessible. Two states for each value of the allowed energy, one for each side, would exist. Actually, because only a finite potential 'hump' is present between the two wells, the two states interact with each other to give two new states. In these states, we cannot say that the nitrogen atom is on one side or the other of the plane, but rather that it has equal probability of being on either side.

The wavefunction describing one of the two states remains unaltered by an interchange of the two positions, while the other wavefunction has its sign changed. The interaction splits the energies of the two new states with the symmetric state being somewhat lower than the antisymmetric. The energy separation between members of a pair increases with increasing pair energy, but for the ammonia molecule it corresponds to frequencies in the microwave range.

Besides these states, which are vibrational in character, the molecule also has rotational states corresponding to rotations around either of two axes, one perpendicular to the hydrogen plane and the other lying in it. The rotational states are identified by the value of the (quantised) angular momenta about each of these axes. Rotation alters the vibrational potential energy curves, stretching the molecule with centrifugal force, so that the separation of each vibrational pair depends on the rotational state[8].

Gordon, Zeiger and Townes, after some alteration of ideas, decided to look at the transition between the lower vibrational pair in the rotational state with three units of angular momentum about each axis, called the 3–3 state. This

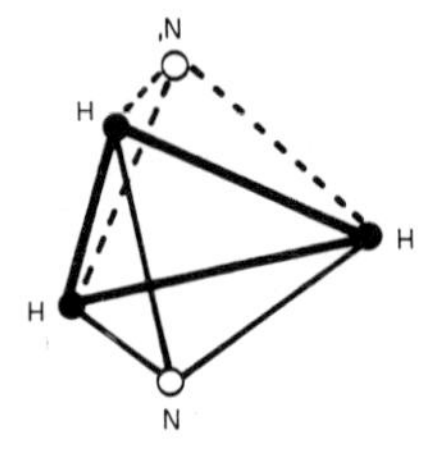

Figure 4.1 The pyramidal structure of the ammonia molecule.

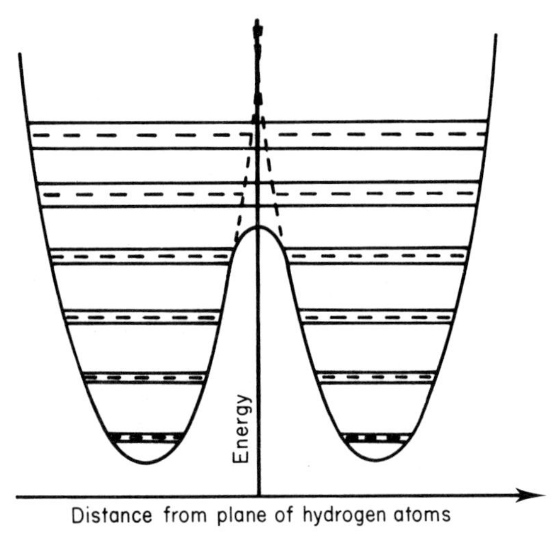

Figure 4.2 Double minima of the potential curve for ammonia which give rise to the inversion splitting of the vibrational levels.

transition has a frequency of 23 870 MHz. [It must be remembered that another important property of the ammonia molecule is that, although it has no permanent electric dipole moment, an applied electric field will induce such a moment in both members of the rotational pair, but opposite in sign. If the field is inhomogeneous, a force will act on the molecule which has opposite direction for both members of the pair.] The project was then to use a strong electrostatic field to obtain a beam of excited ammonia molecules and to focus this beam through a small hole in a box or cavity which had been tuned to resonance at exactly 24 000 MHz.

By concentrating the excited ammonia molecules Townes was hoping to obtain either population inversion in them or else a surplus of excited molecules. After having reached the condition of population inversion, each excited molecule, on decaying into its fundamental state, would trigger other molecules to relax. Therefore the cavity would emit coherent radiation at 24 000 MHz. It is to the credit of Townes that he clearly understood at the outset the need for a resonant cavity with which to couple the radiation to the excited medium.

Townes said that he felt responsible, particularly for Gordon; 'I'm not sure it will work, but there are other things we can do with it if it doesn't'. In fact, he told Gordon that if the method did not work he could use the set-up to investigate the microwave absorption spectrum of ammonia. Gordon thus worked simultaneously on both experiments. In this manner, he was able to study the hyperfine structure of ammonia (the interaction between the electron and the nucleus) with an accuracy higher than had been possible before[9].

The resulting quarterly reports on this laboratory work had a certain amount of circulation among scientists who were interested in microwave physics. The first published mention of this maser project appears, under the names of Zeiger and Gordon, in a report of 31 December 1951, headed *Molecular Beam Oscillator*. Preliminary calculations on the design of a molecular beam oscillator were reported there. A description was given of the essential elements of the oscillator: a molecular beam source; a deflecting region for separating an excited state of the beam from a ground state; a resonant cavity, tuned to the frequency of transition from that excited state to the ground state by induced emission; and a detector for observing the radiation emitted from such transitions.

This projected oscillator was intended for use in the long-wavelength infrared ($\lambda \sim 0.5$ mm), and the transition considered was $J=2$, $K=1$, $M=2 \rightarrow J=1$, $K=1$, $M=1$ in ND_3, which has a transition energy of 20.55 cm^{-1}. After some considerations concerning the focusing system, the total beam flux in the upper state entering the cavity was calculated to be 6×10^{12} molecules per second which, if all these were to undergo induced transitions to the ground state, would deliver approximately 2.4×10^{-9} W of power to the cavity. This was estimated to be sufficient to be detected by a Golay cell.

A theoretical calculation was also made of Q for a tuned cylindrical cavity. For a cavity 1 cm in diameter and 1 cm in length, at liquid-air temperature, and tuned to 20.55 cm^{-1}, the calculated Q was 1.5×10^5. The Q which had been calculated as necessary to maintain oscillations for 2.4×10^{-9} W input was 1×10^5. The authors concluded: 'On the basis of these calculations, it therefore seems only barely possible that oscillations will be sustained in the cavity'.

In the following quarterly report the goal of the project was changed. Operation was now in the K band, using the ammonia inversion transition $J=3$, $K=3$ at 1.25 cm wavelength. Subsequent reports concern the details of the vacuum system; of the focusing equipment; of the resonant cavity; and of the microwave resonator.

For two years the Townes group worked on. At about this time two friends called at the laboratory and tried to insist that Townes stop this nonsense and the wastage of government money, for Townes had by then spent about \$30 000 under a Joint Services grant administered by the Signal Corps, the US Office of Naval Research and the Air Force.

Finally, one day in 1953, Jim Gordon rushed into a spectroscopy seminar that Townes was attending crying: it works! The story goes that Townes, Gordon and the other students (Zeiger had by this time left Columbia to go to the Lincoln Laboratory and T C Wang had replaced him) went to a restaurant both to celebrate and to find a Latin or Greek name for the new device, the latter without success. Only a few days later, with the help of some of the students, they coined the acronym MASER: Microwave Amplification by Stimulated Emission of Radiation. This name appears in the title of a paper in *Physical Review*[10] and its meaning was fully spelt out in a subsequent paper by K Shimoda *et al*[11] (Detractors reread it as Means of Acquiring Support for Expensive Research!)

The first mention of the operation of the oscillator was in a report of 30 January 1954, in nearly the same form in which it was published in a letter to *Physical Review*[12]:

> A block diagram of the apparatus is shown in figure 1†. A beam of ammonia molecules emerges from the source and enters a system of focusing electrodes. These electrodes establish a quadrupolar cylindrical electrostatic field whose axis is in the direction of the beam. Of the inversion levels, the upper states experience a radial inward (focusing) force, while the lower states see a radial outward force. The molecules arriving at the cavity are then virtually all in the upper states. Transitions are induced in the cavity, resulting in a change in the cavity power level when the beam of molecules is present. Power of varying frequency is transmitted through the cavity, and an emission line is seen when the klystron frequency goes through the molecular transition frequency.
>
> If the power emitted from the beam is enough to maintain the field strength in the cavity at a sufficiently high level to induce transitions in the following beam, the self-sustained oscillations will result. Such oscillations have been produced. Although the power level has not yet been directly measured it is estimated at about 10^{-8} W. The frequency stability of the oscillation promises to compare favourably with that of other possible varieties of 'atomic clocks'.
>
> Under conditions such that oscillations are not maintained, the device acts like an amplifier of microwave power near a molecular resonance. Such an amplifier may have a noise figure very near to unity.

† The figure reproduced here (our figure 4.3) is in fact figure 1 from the reference in note 10, but the two figures are substantially identical.

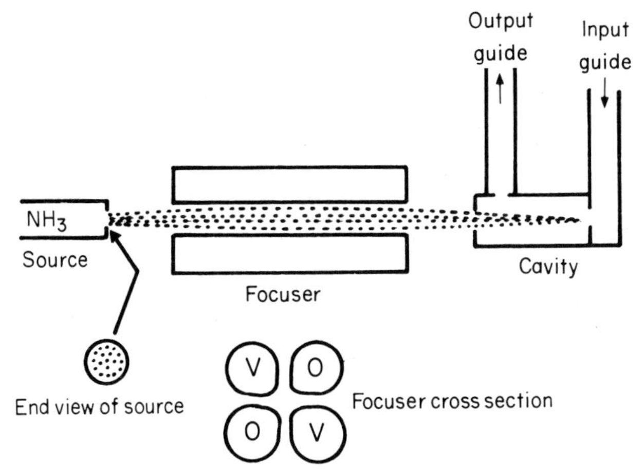

Figure 4.3 Simplified diagram of the essential parts of the ammonia maser[10].

High resolution is obtained with the apparatus by utilising the directivity of the molecules in the beam. A cylindrical copper cavity was used, operating in the TE011 mode. The molecules, which travel parallel to the axis of the cylinder, then see a field which varies in amplitude as $\sin(\pi x/L)$, where x varies from 0 to L. In particular, a molecule travelling with velocity v sees a field varying with time as $\sin(\pi vt/L)\sin(\Omega t)$, where Ω is the frequency of the RF field in the cavity. A Fourier analysis of this field, which the molecule sees from $t=0$ to $t=L$, gives a frequency distribution whose amplitude drops to 0.707 of its maximum at points separated by a $\Delta\nu$ of 1.2 v/L. The cavity used was twelve centimetres long, and the most probable velocity of ammonia molecules in a beam at room temperature is 4×10^4 cm s^{-1}. Since the transition probability is proportional to the square of the field amplitude, the resulting line should have a total width at half-maximum given by the above expression, which in the present case is 4 kc/sec [4 kHz]. The observed linewidth of 6–8 kc/sec [6–8 kHz] is close to this value.

Consideration as to the use of the device as a spectrometer was given later. In the subsequent paper sent to *Physical Review* exactly one year later[10] more particulars are given.

The electrodes of the focuser were arranged as shown in figure 1 [our figure 4.3]. High voltage is applied to the two electrodes marked V, while the other two are kept at ground. Paul *et al*† have used similar magnetic pole arrangements for the focusing of atomic beams.

† H Friedburg and W Paul, *Naturwiss.* **38**, 159 (1951); H G Bennewitz and W Paul, *Z. Phys.* **139**, 489 (1954).

The electric field increases with radial distance from the central axis of the focusing electrodes. Because the force on molecules is $F = -\partial W/\partial r$ and because the molecules in the upper state increase in energy with increasing field, these molecules are forced towards the centre of the focuser. The lower-energy molecules, on the other hand, are turned away from the central axis of the focuser.

The cavity, in practice, worked as the positive reactance in an electronic circuit, amplifying the radiation at its resonant frequency. The basic requirement for microwave generation was therefore to produce a positive feedback through some resonant circuit which ensured that the gain in energy received by the wave through stimulated emission was greater than the circuit losses[11].

Let us consider a microwave resonant cavity with conducting walls, volume V and a quality factor Q; this last quantity is defined as the ratio between the incident power and the power loss P_R due to wall resistance

$$Q = \overline{E^2}V\nu/4P_R, \tag{4.7}$$

$\overline{E^2}$ being the mean square value (for that volume) of the electric field of the mode and ν its frequency. If a molecule in an excited state is put into an electric field E at resonance (when the frequency ν coincides with the energy difference between the levels divided by h) the emitted power is

$$P_e = (\overline{E^2}\mu^2/\hbar^2)(h\nu/3\Delta\nu), \tag{4.8}$$

where μ is the matrix element for the transition and $\Delta\nu$ is the half linewidth of the resonance. Therefore for N_b molecules, in the upper state, and N_a in the lower state, the power given by the field to the cavity is

$$P_e = (N_b - N_a)(\overline{E^2}\mu^2/\hbar^2)(h\nu/3\Delta\nu). \tag{4.9}$$

If molecules are distributed uniformly in the volume it follows that

$$(N_b - N_a)(\overline{E^2}\mu^2/\hbar^2)(h\nu/3\Delta\nu) \geqslant \overline{E^2}V\nu/4Q, \tag{4.10}$$

from which the threshold condition for the onset of oscillations in the cavity is derived as

$$(N_b - N_a) \geqslant 3hV\Delta\nu/16\pi^2 Q\mu^2. \tag{4.11}$$

Townes observed immediately that[10]:

> Associated with the power emitted from the beam is an anomalous dispersion.... These two effects can be considered at the same time by thinking of the beam as a polarisable medium introduced into the cavity whose average electric susceptibility is given by $\chi = \chi' + i\chi''$. The power

emitted from the beam is

$$P = 8\pi^2 \nu_B W \chi'', \tag{4.12}$$

where W is the energy stored in the cavity.

The connection between the imaginary part of the susceptibility and the absorbed or emitted power had already been considered by both Bloch and Bloembergen and Purcell and Pound[13, 14], and various treatments of the maser were achieved by having recourse to circuit analogies[15].

The principal property of the maser is its extremely low noise, both as an amplifier and as an oscillator. It is therefore able to amplify extremely low-level signals. When it is used as an oscillator it is able to generate monochromatic radiation with very good frequency stability. The maser monochromaticity is described quantitatively by the halfwidth, $\delta\nu$, of its emission spectrum. Townes[10] (using a linear approximation) gave the expression

$$\delta\nu = 4\pi kT(\Delta\nu)^2/P, \tag{4.13}$$

where $\Delta\nu$ is the halfwidth of the spectral line. This is the theoretical lower limit, although that is not clearly stated. A different expression which did take into account the non-linear behaviour of the maser was given later by Shimoda, Wang and Townes[11].

At operational temperatures and at the powers actually used, $\delta\nu$ is several orders of magnitudes smaller than $\Delta\nu$. The noise generated in a maser has been the object of a number of studies and measurements[16]. The centre frequency, ν_0, of the oscillation was derived in these studies as

$$\nu_0 = \nu_B + (\Delta\nu_B/\Delta\nu_c)(\nu_c - \nu_B), \tag{4.14}$$

where $\Delta\nu_c$ and $\Delta\nu_B$ are the halfwidths of the cavity mode and of the molecular emission line, respectively, and $\nu_c - \nu_B$ is the difference between the cavity resonant frequency ν_c and the line frequency ν_B. (It is worth observing that the coherence of the radiation was never explicitly mentioned!)

The only work published before the successful operation of the maser was by Townes in 1953[17] but it was essentially a translation of the forthcoming paper in *Physical Review*[12]. Mention of the work on stimulated emission had also been made in May 1951 at a *Symposium on Submillimeter Waves* at the University of Illinois by A H Nethercot of Columbia[5] on behalf of Townes.

It was immediately realised that one important application of molecular beam masers would be molecular spectroscopy. Molecular beams had already been considered by gas spectroscopists in the early 1950s; however, the basic problem had been that, as a consequence of beam formation, the resultant density of molecules in the spectrometer cell was very low. For beams where the molecules may be assumed to be in or very near thermal equilibrium, the absorption and emission processes in the presence of external radiation nearly balance out against a small net absorption which occurs because more molecules are in the lower energy state. If all the molecules in the lower state are removed, as in maser operation, then the magnitude of the signal can be enhanced by a factor of $kT/h\nu$ over its thermal equilibrium value which, at microwave frequencies and room temperature, is more than two orders of magnitude. Although the first spectroscopic study had already been undertaken by Gordon[9] in 1951 the full potential of the method was not realised until the early 1970s[18].

4.4 Basov and Prokhorov and the Soviet approach to the maser

There is a patent filed on 18 June 1951 by V A Fabrikant of the Moscow Power Institute, together with some of his students, which was only published in 1959, entitled *A method for the amplification of electromagnetic radiation (ultra-violet, visible, infrared, radio wavebands)*[19]. The title is so general as to cover nearly anything connected with maser or laser action. However, the work seems to have been principally geared towards lasers, and so it will be considered later.

The idea of using a gas as a molecular amplifier – or *molecular generator* as they called it – came to Basov and Prokhorov at the Lebedev Institute, Moscow. They published a theoretical paper[21] a few months before the paper by Townes[12] appeared in *Physical Review*.

Alexander Mikhailovich Prokhorov was born on 11 July 1916 in Atherton, Australia, to the family of a revolutionary worker who emigrated there from exile in Siberia in 1911. Prokhorov's family returned to the Soviet Union in 1923. In 1939 he graduated from Leningrad University and went to work at the Lebedev Institute of Physics (FIAN) Moscow, one of the most prestigious research institutes of the USSR Academy of Sciences. He started his scientific work in 1939, studying the propagation of radio waves over the Earth's surface.

During the Second World War he was wounded twice and returned to the Institute in 1944. After the war, following a suggestion by V I Veksler, he demonstrated experimentally in his doctoral thesis that the synchrotron can be used as a source of coherent electromagnetic oscillations in the centimetre waveband. He went on to head a group of young research workers (amongst whom was Basov) working on radio wave spectroscopy.

After his work in the field of masers and lasers, which we shall consider in a moment, he was in 1960 elected a Corresponding Member of the USSR Academy of Sciences and in 1966 he became a Full Member. For his researches he was awarded the title of Hero of Socialist Labour, the Lenin prize and, in 1964, together with Basov and Townes, the Nobel Prize.

Nikolai Gennadievich Basov was born on 14 December 1922. At the start of the Second World War he graduated from a secondary school in Voronezh, and enlisted. He was sent first to Kuibyshev and then to the Kiev school of Military Medicine, from which he graduated in 1943 with the rank of lieutenant in the medical corps. His service began in the chemical warfare defence forces and then continued at the front. A little after the end of the war, following his return from Germany, he realised his dream of studying physics while still in the Soviet Army. He enrolled at the Moscow Institute of Mechanics (now of Physics Engineering). Exactly 20 years later he was elected to the USSR Academy of Sciences.

In 1948 Basov began work as a laboratory assistant in the Oscillation Laboratory of the Lebedev Institute of Physics. The laboratory was headed at that time by M A Leontovich and Basov later became an engineer there. Later still, in the early 1950s, a group of young physicists under the leadership of A M Prokhorov began work there on the study of molecular spectroscopy. This marked the start of many years of fruitful collaboration in the fields of masers and lasers between Basov and Prokhorov, as will be seen later.

This group of researchers was interested in both rotational and vibrational molecular spectroscopy. The possibility of using microwave absorption spectra to produce frequency and time standards was also investigated. The operational accuracy of microwave frequency standards is determined by the resolving power of the radio spectroscope. This in turn depends exclusively on the width of the absorption line itself. An effective way of narrowing down the absorption line was found to be to use spectroscopes operating in conjunction with molecular beams. However, the capabilities of molecular spectroscopes were strongly limited by the low intensity of the observed lines, which in turn was determined by the small population differences of the quantum transition investigated at microwave frequency. The idea that it was possible to increase the sensitivity of the spectroscope appreciably by artificially varying the populations in the levels arose at this stage of their work. In a review paper written in 1955[20] they claim that they had already pointed out the theoretical possibility of constructing a device producing microwaves by using stimulated emission at an *All Union Conference on Radio Spectroscopy* in May 1952. However, their first written paper was not published until October 1954[21]. It contained a detailed theoretical study of the use of molecular beams in microwave spectroscopy. In this the group showed that molecules of the same kind, present in a beam containing molecules in different energy states, can be separated one from the other by letting the beam pass through a non-uniform electric field. Molecules in a pre-selected energy state can then be sent into a microwave resonator where absorption or amplification takes place. The paper contained detailed calculations on the role of the relevant physical parameters, the effects of linewidth, of cavity dimensions, and so on.

Calculations, applied to the rotational spectrum of CsF, had indicated that the required cavity factor Q for a signal generator could not be obtained in this case. However, the quantitative conditions for the operation of a microwave amplifier and of a generator – called by them a molecular generator – were given.

In a subsequent paper, submitted for publication on 1 November 1954[22], Basov and Prokhorov proposed (shortly before Bloembergen[25]) a three-level method, considering in this case a gas to be the active medium.

They examined two possible three-level schemes (figure 4.4). With reference to the first method, figure 4(*a*), at equilibrium

$$n_1 < n_2 < n_3. \tag{4.15}$$

The authors considered the use of a strong radiation source $\nu_{31} = \nu_{ex}$ such that

$$\nu_{31} = (E_1 - E_3)/h, \tag{4.16}$$

in order to saturate the transition and to have $n_1 \simeq n_3$. The reduction of population on level 3 and the increase on level 1 was indicated as producing a situation of inverted population in which

$$n_1 > n_2. \tag{4.17}$$

In figure 4.4(*b*), on the other hand, the case is shown in which a decrease of molecules on level 2 is obtained, which thus makes possible inversion between level 1 (near enough to level 2 so as to be populated thermally) and level 2.

Although these systems could be used as amplifiers, they would lack the tunability of the paramagnetic system proposed soon afterwards by Bloembergen. Moreover, all discussion of relaxation mechanisms was omitted. The methods considered were, in many cases, based on the various rotational and vibrational levels of the molecules, and none of these has ever worked.

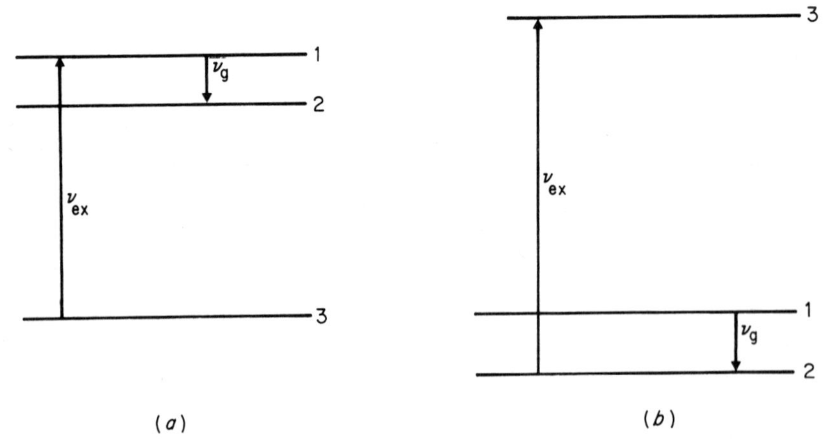

Figure 4.4 The three-level configurations of Basov and Prokhorov in gas molecules.

Basov and Prokhorov also developed a theory of the molecular oscillator using a semiclassical approach. They began by considering a cavity filled with a medium, with negative losses in the neighbourhood of a frequency ω_g. The active medium was characterised by a complex dielectric constant[20, 23]

$$\epsilon = \epsilon' - i\epsilon'' = 1 + 4\pi N\chi, \tag{4.18}$$

where χ is the molecular polarisability and N is the density of molecules in the cavity.

The value of χ was written as follows:

$$\chi = (|m_{zn}^m|^2 \omega/\hbar\omega_g)\{[\omega_g - \omega - (i/\tau)](\rho^0 - \rho_n^0)/[(\omega - \omega_g)^2 + (1/\tau^2) + |m_{zn}^m|^2 (E^2/\hbar^2)]\}, \tag{4.19}$$

where ω_g is the central frequency of the atomic line; l is the cavity length; $\bar{v}$ is the mean velocity of molecules; $\tau = l/\bar{v}$ is the mean time for molecules to travel through the field; ρ_k^0 is the probability of molecules being at the level k at the moment of arriving at the field inside the cavity; and m_{zn}^m is a dipole matrix element.

If N_0 is the number of active molecules (the difference between the numbers of molecules in the upper and lower levels) travelling through the cavity cross section S every second, then

$$N = N_0/S\bar{v}. \tag{4.20}$$

Substituting equations (4.19) and (4.20) into equation (4.18) one obtains the expressions for ϵ' and ϵ''

$$\epsilon' = [1 - A\gamma(\omega/\omega_g)(\omega - \omega_g)\tau]/[(\omega - \omega_g)^2 + (1/\tau^2) + \gamma|E|^2], \tag{4.21}$$
$$\epsilon'' = [-A\gamma(\omega/\omega_g)]/[(\omega - \omega_g)^2 + (1/\tau^2) + \gamma|E|^2],$$

where

$$A = 4\pi\hbar N_0/Sl, \qquad \gamma = |m_{zn}^m|^2/\hbar^2. \tag{4.22}$$

By assuming that the electric field intensity in the cavity section is uniform they then wrote

$$(d^2E/dt^2) + (\omega_0/Q)(dE/dt) + (\omega_0^2/\epsilon)E = 0, \tag{4.23}$$

where ω_0 is the natural frequency of the cavity without molecules.

For a stationary condition

$$E = E_0 \exp(i\omega t). \tag{4.24}$$

Substituting the expressions for ϵ and E into equation (4.23) and taking the real and imaginary parts equal to zero, they obtained two equations for E_0 and ω:

$$-\omega^2 + \omega_0^2\epsilon'/[(\epsilon')^2 + (\epsilon'')^2] = 0,$$
$$\omega/Q + \omega_0\epsilon''/[(\epsilon')^2 + (\epsilon'')^2] = 0, \qquad (4.25)$$

If $\omega_0 = \omega_g \simeq \omega$ and $E \to 0$, the condition for self-excitation is obtained as

$$z = (4\pi N_0/Sl\hbar)|m_{z_n}^m|^2 Q\tau^2 \to 1. \qquad (4.26)$$

Otherwise, if $z \gg 1$, the approximate expression for a stationary amplitude oscillation is

$$E_0^2 = (4\pi N_0 \hbar\omega_g/Sl)\,Q, \qquad (4.27)$$

and the maximum power surrendered by molecules to the cavity is equal to $1/2N_0\hbar\omega_g$.

The frequency

$$\omega \simeq \omega_g[1-(2Q/\omega_g\omega_0\tau)(\omega_0-\omega_g)], \qquad (4.28)$$

results if

$$(\omega_0-\omega_g)/\omega_0 \ll 1. \qquad (4.29)$$

If the self-oscillation condition in the molecular oscillator is not fulfilled, the device can be used as a power amplifier, in which case the following equation is valid:

$$(\mathrm{d}^2E/\mathrm{d}t^2)+(\omega_0/Q)(\mathrm{d}E/\mathrm{d}t)+(\omega_0^2/\epsilon)\,E = B\omega^2 \exp(\mathrm{i}\omega t), \qquad (4.30)$$

where B is the amplitude of the external force. The solution is

$$E = A \exp(\mathrm{i}\omega t) \qquad (4.31)$$

with, at the resonance condition,

$$A \simeq B/[(1/Q)-(4\pi N_0|m_{z_n}^m|^2\tau^2/Sl\hbar)]. \qquad (4.32)$$

Basov worked actively in the new field of *quantum radiophysics*, as it was called in the Soviet Union, and for his doctoral thesis assembled the first Soviet maser a few months after Townes[24].

4.5 The three-level solid-state maser

In 1956 Nicolaas Bloembergen suggested an alternative approach with paramagnetic materials[25] by using three levels instead of two[26]. Discussion on the use of paramagnetic ions had previously been put forward by M W P Strandberg a microwave spectroscopist at MIT[27] who gave a seminar on this theme at which Bloembergen participated.

Bloembergen was born in Dordrecht, The Netherlands on 11 March 1920. He studied under L S Orstein and L Rosenfeld and received the Phil. Cand. and Phil. Drs. degrees from the University of Utrecht in 1941 and 1943, respectively, during the German occupation of the Netherlands. Then he escaped to the USA and went to Harvard where he arrived six weeks after Purcell, Torrey and Pound had detected nuclear magnetic resonance. They were busy writing a volume for the MIT Radiation Laboratory series on microwave techniques, and the young Bloembergen was accepted as a graduate assistant and asked to develop the early NMR apparatus, so he started to study nuclear magnetic resonance and in the meantime attended lectures by J Schwinger, J H van Vleck, E C Kemble and others.

He returned to the Netherlands for a short period after the war and pursued his research in a postdoctoral position at the Kamerlingh Onnes Laboratory in 1947–8 at the invitation of C J Gorter who was a visiting professor at Harvard during the summer of 1947. In 1948 he received a PhD at Leyden University with a thesis on nuclear paramagnetic relaxation which was subsequently published as a short book†. He then went back to Harvard and joined Purcell and Pound in work on magnetic resonance to which he made important contributions, some of which were referred to in the preceding chapter. In 1951 he became an Associate Professor and in 1957 Gordon McKay Professor of Applied Physics at Harvard, where he has been the Gerbard Gade University Professor since 1980. His important research in the fields of nuclear magnetic resonance, masers and non-linear optics led to the award of the 1981 Nobel Prize for Physics (an award shared with Schawlow and Siegbahn).

Bloembergen proposed a tunable maser utilising Zeeman levels in a paramagnetic material. He was influenced by the work of Overhauser and wrote[25]

> Attention is called to the usefulness of power saturation of one transition in a multiple energy level system to obtain a change of sign of the population difference between another pair of levels. A variation in level populations obtained in this manner has been demonstrated by Pound‡. Such effects have since acquired wide recognition through the work of Overhauser§.

To understand Bloembergen's proposal let us recall that a paramagnetic material with spin $\frac{1}{2}$ has two states for each electronic configuration. In a static magnetic field H (in oersted) these levels are separated by an energy difference that corresponds to a frequency (in MHz)

$$\nu = 2.8H. \tag{4.33}$$

† *Nuclear Magnetic Relaxation* (W A Benjamin: New York, 1961).
‡ R V Pound, *Phys. Rev.* **79**, 685 (1950).
§ A W Overhauser, *Phys. Rev.* **92**, 411 (1953).

Atoms or ions with n unpaired electrons have $n+1$ such levels, which are degenerate in the absence of an external field, but which are separated in a crystalline field or in some other external field. The solid-state maser proposed by Bloembergen used these magnetically separable levels. Tuning of this maser is obtained by varying the strength of the external magnetic field.

Let us now consider a material having three relevant, unequally spaced energy levels (figure 4.5). Some paramagnetic ions in single crystals, usually immersed in a magnetic field, have levels separated by transitions at microwave frequencies. At thermal equilibrium the populations obey the conditions

$$n_{10} > n_{20} > n_{30} \tag{4.34}$$

but, being in the microwave region, where generally $h\nu < kT$, all three levels are substantially populated.

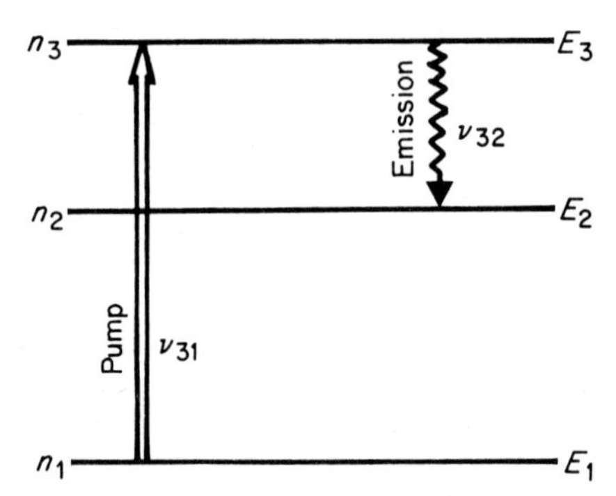

Figure 4.5 The three-level configuration of a paramagnetic material.

The system is now subjected to a strong *pumping* radiation at frequency ν_{31} such as to induce transitions between levels 1 and 3. Because, initially, more atoms are in the fundamental level 1, the system will absorb energy, populating level 3 at the expense of level 1. The net effect is that the populations n_1 and n_3 tend to become equal. An exact calculation of the populations of each of the three energy levels in a stationary state was made by Bloembergen as follows. Let us assume $E_3 > E_2 > E_1$ and put

$$h\nu_{31} = E_3 - E_1 \qquad h\nu_{32} = E_3 - E_2 \qquad h\nu_{21} = E_2 - E_1. \tag{4.35}$$

The transition probabilities between these spin levels under the influence of the thermal motion of the heat reservoir (lattice) are

$$\begin{aligned} w_{12} &= w_{21} \exp(-h\nu_{21}/kT) \\ w_{13} &= w_{31} \exp(-h\nu_{31}/kT) \\ w_{23} &= w_{32} \exp(-h\nu_{32}/kT). \end{aligned} \tag{4.36}$$

The w correspond to the inverse of the spin-lattice relaxation times. We denote the transition probability caused by a large saturating field $H(\nu_{31})$ of frequency ν_{31} by W_{31}. Let a relatively small signal of frequency ν_{32} cause transitions between levels two and three at a rate W_{32}. The number of spins occupying the three levels, n_1, n_2 and n_3, satisfy the conservation law

$$n_1 + n_2 + n_3 = N. \tag{4.37}$$

For $h\nu_{32}/kT \ll 1$ the populations obey the equations

$$\begin{aligned} dn_3/dt &= w_{13}[n_1 - n_3 - (N/3)(h\nu_{31}/kT)] + w_{23}[n_2 - n_3 - (N/3)(h\nu_{32}/kT)] \\ &\quad + W_{31}(n_1 - n_3) + W_{32}(n_2 - n_3) \\ dn_2/dt &= w_{23}[n_3 - n_2 + (N/3)(h\nu_{32}/kT)] + w_{21}[n_1 - n_2 - (N/3)(h\nu_{21}/kT)] \\ &\quad + W_{32}(n_3 - n_2) \\ dn_1/dt &= w_{13}[n_3 - n_1 + (N/3)(h\nu_{31}/kT)] + w_{21}[n_2 - n_1 + (N/3)(h\nu_{21}/kT)] \\ &\quad - W_{31}(n_1 - n_2). \end{aligned} \tag{4.38}$$

In the steady state the left-hand sides are zero. If the saturating field at frequency ν_{31} is very large, $W_{31} \gg W_{32}$ and all w_{ij}, the following solution is obtained:

$$n_1 - n_2 = n_3 - n_2 = hN/3kT\,(-w_{23}\nu_{32} + w_{21}\nu_{21})/(w_{23} + w_{12} + W_{32}). \tag{4.39}$$

This population difference will be positive, corresponding to negative absorption or stimulated emission at the frequency ν_{32} if

$$w_{21}\nu_{21} > w_{32}\nu_{32}. \tag{4.40}$$

The power emitted by the magnetic specimen is also calculated as

$$P_{\text{magn}} = (Nh^2\nu_{32}/3kT)\,[(w_{21}\nu_{21} - w_{32}\nu_{32})\,W_{32}]/(w_{32} + w_{12} + W_{32}). \tag{4.41}$$

The choice of a paramagnetic substance that Bloembergen made was dependent on the existence both of suitable energy levels and of matrix elements of the magnetic moment operator between the various spin levels. It is essential that all off-diagonal elements between the three spin levels under consideration be non-vanishing. This can be achieved by putting a paramagnetic salt in a magnetic field in some suitable way. In this way the states with magnetic quantum numbers m_s are mixed up. The essential role played by relaxation is evident here.

Bloembergen also considered some possible materials, mentioning nickel fluorosilicate and gadolinium ethyl sulphate. The three-level maser was the first maser to offer the practical advantage of continuously tunable amplification with a reasonable bandwidth at microwave frequencies, still maintaining the principal characteristic of a maser (an extremely low noise figure).

Unfortunately the Harvard group, being interested (for astronomical purposes) in a devicc working on the interstellar H line at 1420 MHz failed to obtain the first successful operation of the three-level maser. This took place the following year at Bell Telephone Laboratories, operated by H E D Scovil, G Feher and H Seidel. They had built the maser using the Gd^{3+} paramagnetic ions in a host lattice of lanthanum ethyl sulphate[28]. A short time later Alan L McWhorter and James W Meyer at MIT Lincoln Laboratory[29] used Cr^{3+} ions in $K_3Co(CN)_6$ to build the first amplifier. The inversion requirement $w_{23}\nu_{23} \gg w_{12}\nu_{12}$ in the Scovil maser was obtained by altering the ratio w_{12}/w_{23} by the introduction of Cr^{3+} into the crystal. This technique of cross-doping was necessary because the energy levels used were such that $\nu_{12} \simeq \nu_{23}$ for the orientation and field chosen. The preliminary results of electron spin relaxation times were published by Feher and Scovil in a preceding letter[30]. While the original ammonia maser was principally useful as a frequency standard or else as a very sensitive detector, the solid-state maser was something which really could be used for communications and radar. It had a larger bandwidth and could be tuned by changing the magnetic field strength.

Not much later, C Kikuchi and his co-workers[31] of Michigan University showed that ruby was a good material for such masers. It was Joseph Geusic who became active in the design and perfection of the ruby maser. He had just gone to Bell Laboratories from Ohio State University where he had written his thesis under J G Daunt dealing, for the first time, with the measurement of microwave resonance in ruby[32]. (Paramagnetic resonance in ruby had already been investigated in the Soviet Union in 1955–6[33].)

During 1957 and 1958 many masers were built in several laboratories, including Harvard[34], by using Cr^{3+} ions in ruby crystals. Rubies were employed in a great number of types of maser with many different characteristics. For example at Bell Laboratories[35] a travelling-wave ruby maser was assembled working below 2 K having a noise temperature that was too low to be measured: it was a most sophisticated laboratory instrument. At Hughes Research Laboratory, on the other hand, a compact lightweight device was built operating at 77 K with a noise temperature of about 93 K[36]; very useful for various applications.

Since 1958 many masers have been built for applications in radioastronomy or else as components in radar receivers[37]. These masers were almost all of the ruby type. A solid-state maser was used by A Penzias and R W Wilson in 1965 to discover the 3K black-body radiation from the Big Bang[38]. Travelling-wave masers were first suggested by H Motz[39] in 1957 and realised and discussed by de Grasse *et al*[35] in 1959.

The use of two levels of a paramagnetic substance in a magnetic field was tried without success by Townes using Ge[40]. He obtained neither amplification nor oscillation conditions. This proposal was based on the use of adiabatic fast passage, as discussed by Bloch in his fundamental paper[13]. In the solution for the components of magnetisation given in equation (3.11), the sign of M_z depends on the sign of $\delta = [H_0(t) - \omega/|\gamma|]/H_1$. The quantity δ is zero at resonance where the resonant field is $H_0 = \omega/|\gamma|$. In the adiabatic fast passage $\delta(t)$ remains constant and then at time t_0 it is quickly increased through resonance (subject to the 'adiabatic' condition $|d\delta/dt| \ll |\gamma H_1|$). Thus its sign changes as it goes rapidly but adiabatically through resonance, and according to equation (3.11(c)) this means the sign M_z has been changed and is now anti-parallel to H_0.

Although the early attempt by Townes[40] was unsuccessful, two years later Feher *et al*[41] obtained maser action due to adiabatic fast passage using the same material proposed by Townes (paramagnetic electrons associated with the P donors in Si). A volume of 0.3 cm^3 of isotopically purified Si^{28} with a phosphorus concentration of 4×10^{16} atoms/cm^3 was used. Radiation at 9000 MHz at 1.2 K was obtained for about 50 μs.

Using neutron irradiated quartz and magnesium oxide, Chester *et al*[42] obtained similar results. Their quartz sample contained 10^{18} spins and the inverted population persisted for 2 μs at 4.2 K. The emission frequency was 9000 MHz and the maser operated as an amplifier of better than 20 dB gain.

A thermodynamic approach to the three-level maser was made in 1959 by Scovil and Schulz-DuBois[43]. Considering figure 4.6, levels 1 and 3 are supposed to be in thermal contact, through a filter passing frequencies in the vicinity of the pumping frequency $\nu_p = \nu_{13}$ and rejecting frequencies in the vicinity of ν_{23} and ν_{21}, with a heat reservoir at temperature T_1. Levels 2 and 3 are in thermal contact with a reservoir at a lower temperature T_0 through a filter which passes frequencies in the vicinity of the idler frequency $\nu_i = \nu_{23}$ but rejects those close to ν_p and the signal frequency $\nu_s = \nu_{21}$. During maser operation, for each quantum $h\nu_p$ supplied by the heat source, the energy $h\nu_i$ is passed to the heat sink, and the efficiency of the system is

$$\eta_M = \nu_s/\nu_p.$$

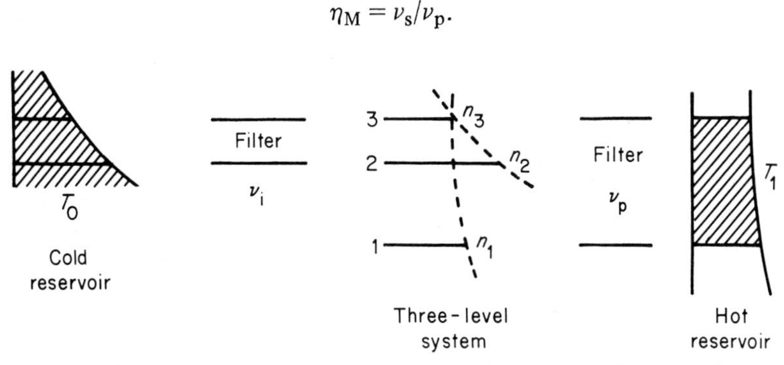

Figure 4.6 Three-level system in thermal contact with two heat reservoirs[43].

From Boltzmann's distribution law we find

$$(n_2/n_1) = (n_2/n_3)(n_3/n_1) = \exp(h\nu_i/kT_0)\exp(-h\nu_p/kT_1)$$
$$= \exp\{(h\nu_s/kT_0)[(\nu_p/\nu_s)(T_1-T_0)/T_1 - 1]\}.$$

In this formula one recognises the maser efficiency η_M and the efficiency of the Carnot cycle $\eta_C = (T_1 - T_0)/T_1$. Using these, the condition for maser action is

$$\eta_M \leqslant \eta_C.$$

Maser efficiency equals that of a Carnot engine if the signal transition is at the verge of inversion $n_2 - n_1 \to +0$ or $T_{sig} \to -\infty$.

4.6 Optically pumped masers

Optical pumping techniques for masers have been receiving attention since 1957[44].

The general idea was to use optical pumping to create a population inversion between some pairs of Zeeman sub-levels of the lower atomic energy level. For various reasons, gaseous systems seemed good candidates for this type of pumping. However, the maser action thereby obtained in gases is inherently very weak because of the low spin density in a gas.

Efforts to observe stimulated emission were unsuccessful[45] until, in 1962, Devor and co-workers[46] operating at 4.2 K made possible maser action in ruby by pumping with a laser. Using a magnetic field at an angle of 67° with respect to the crystalline c axis, they split the $\bar{E}$ and $^4A_2$ states as shown in figure 4.7 (which is figure 2(a) of note 46). At 6700 Oe, the $+\frac{1}{2}(\bar{E}) \to +\frac{1}{2}(^4A_2)$ transition matched the $\bar{E} \to \pm\frac{1}{2}(^4A_2)$ component of the pumping ruby laser. Amplification of microwave power was obtained at 22.4 GHz corresponding to the $+\frac{3}{2}(^4A_2) \to +\frac{1}{2}(^4A_2)$ transition.

The general theory was developed simultaneously by Hsu and Tittel[47], who considered the three-level configuration shown in figure 4.8 (which is figure 1 of note 47). The two lower energy levels 1 and 2 both belonging to the ground state, are separated by a microwave transition. The third level 3 is separated from the other two by an optical transition.

The condition for stimulated emission between levels 2 and 1 was obtained (by assuming saturation for the pump transition between levels 1 and 3) as

$$\Delta N/N_1 = (N_2 - N_1)/N_1 = [(w_{32}/w_{21}) - (h\nu_{12}/kT)]\,[1 + (\bar{W}_{21}/w_{21})]^{-1} > 0, \tag{4.42}$$

or

$$(w_{32}/w_{21}) > (h\nu_{12}/kT). \tag{4.43}$$

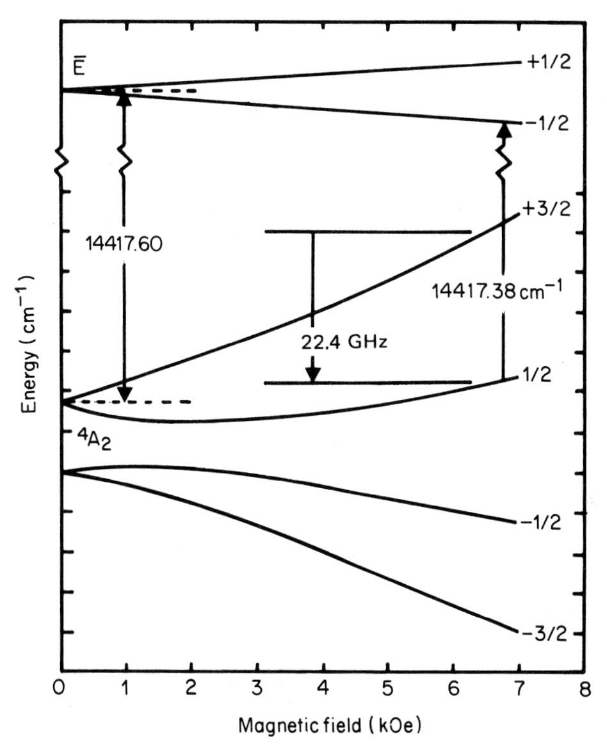

Figure 4.7 Zeeman structure of Cr^{3+} in ruby. The crystalline axis is oriented at 67° with respect to the magnetic field. (From D P Devor, I J D'Haenens and C K Asawa, *Phys. Rev. Lett.* **8**, 432 (1962).

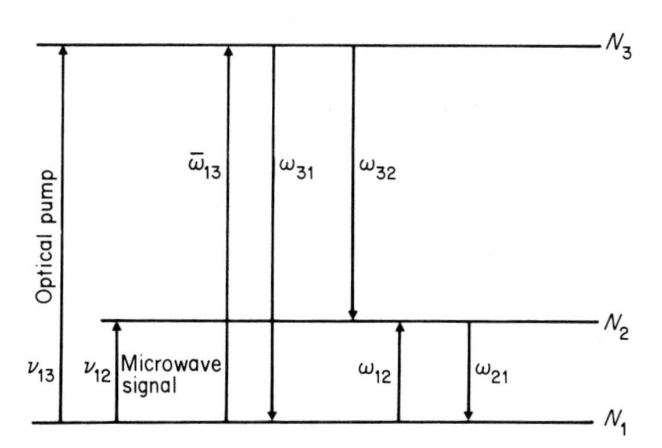

Figure 4.8 Three-level configuration used by Hsu and Tittel for discussing optical pumping of a maser. (From note 47; Copyright © 1963 IEEE.)

Comparing equation (4.43) with the equivalent condition derived by Bloembergen in equation (4.40) one sees that the difference lies in the fact that the former is valid when $h\nu_{32} \gg kT$, and the latter holds for $h\nu_{32} \ll kT$.

From equation (4.43) a limiting signal frequency can be defined as

$$\nu_{12}^{0} = (w_{32}/w_{21})(kT/h). \tag{4.44}$$

Above this frequency the maser ceases to function as a useful device, even with an infinite pumping power.

The excess noise temperature (T_{ex}) was also calculated as a function of operating temperature (T) and the following approximate expression was found

$$T_{ex} \simeq T[(\nu_{12}^{0}/\nu_{12}) - 1]^{-1},$$

showing that the excess noise temperature can be much lower than the operating temperature, which is a particular advantage of this kind of maser.

Almost simultaneously and independently Ready and Chen considered the possibility of optical pumping of masers using the ruby, in a short note in *Proc. IRE*[48], proposing essentially the same scheme as Devor *et al*[46].

4.7 Maser action in nature

An important advance in the science of radioastronomy some years ago was the observation of radio emission at 1420 MHz originating from the gaseous H interstellar clouds. This particular radio emission represents the spontaneous emission, or what could be called 'microwave atomic fluorescence', of a particular transition in atomic hydrogen, coming from hydrogen atoms at thermal equilibrium at a fairly low temperature (less than 100 K). As such it has none of the characteristics of maser amplification but only the usual features of fluorescent emission.

However, some years later, in 1965, from the Radio Astronomy Group led by Professor H Weaver at Berkeley, California[49] came news of surprising observations of radio emission at about 1670 MHz coming from OH molecules located near some stars. This emission, in reality, consisted of four known OH transitions at 1612, 1665, 1667 and 1720 MHz. If emission on these lines had originated spontaneously, the four lines would have had intensities in the ratios 1:5:9:1 as predicted by the well known transition intensities for the four transitions. However, the observed intensity ratios were quite different and changed fairly quickly with time (with a time scale of months). The emission frequency profile of each line was not a smooth one, but sometimes contained very narrow spectral components. The linewidths were such that the temperature of the source would have to be lower than 50 K to have a Doppler broadening as small as that observed. At the same time the emission intensity was so strong as to correspond to a source temperature of 10^{12}–10^{13} K, and the emission apparently originated either from extremely narrow point sources or else appeared in the form of highly directional beams. The different spectral lines were also often polarised in a well defined way, with polarisations which changed over time and were generally different from those which would be expected to result from spontaneous emission.

The only reasonable explanation of these results was that such radiations originated by spontaneous emission in some part of an OH cloud and were then strongly and selectively amplified by a maser amplifier while they were going through other regions of the OH cloud.

Such amplification could explain the anomalous intensity ratios and the high brightness and directionality of the emission. Appropriate Zeeman splittings due to intergalactic magnetic fields could also explain the properties of polarisation.

It also seems reasonable that the maser gain properties can change quickly with time, although changes both in the total quantity of OH and in the associated spontaneous emission should not be expected on the time scales observed. The pumping mechanism responsible for the apparent population inversion is not yet clear: a number of suggestions centre around the optical selective pumping of OH molecules by strong stellar ultraviolet radiation – other possibilities are collisions with electrons, and chemical reactions.

The observed emissions all come from OH clouds near to very intense stars where there is no shortage of radiation or of other energy mechanisms which could excite OH molecules and cause the population inversion. Predictions on the possibility of having negative absorption in radioastronomy date back as far as 1958[51]. In 1968 other substances were found to emit similarly, and today more than 36 molecules and nearly 200 transitions are known[50].

The celestial masers are today believed to exist in regions where stars are forming and can give us more information about their formation process.

4.8 Other proposed systems

In 1958 A Javan[52] proposed a method of obtaining amplification using a nonlinear two-photon process with the kind of parametric amplifier which uses Raman scattering and in which it is not necessary to have population inversion. Javan's work was read at a meeting of the American Physical Society and today only a short summary exists. Weber in his review paper[26] makes mention of it on page 692.

Very little attention has been paid in the past to important work by Motz, who was a pioneer in proposing the production of energy in the millimetre to infrared portion of the spectrum, through a mechanism which is today employed in the so-called 'free-electron beam lasers', and which is used for microwave production in a number of devices (gyrotrons, omichrons, etc)[53].

As early as 1951, Motz[54], then at the Microwave Laboratory, Stanford University, California, described the use of an undulator for the production of millimetre and submillimetre waves by means of a relativistic beam of electrons passing through the magnetic field produced by an array of magnets[55] with alternating polarity (figure 4.9). With suitable pole pieces, the magnetic field near the axis is approximately sinusoidal, the spatial period of the sine wave being Λ. An observer attached to the electron sees the magnet structure contracted by the Lorentz factor $\gamma = (1-\beta^2)^{-1/2}$ where $\beta = v/c$, v being the velocity of electrons. For such an observer the wavelength of the electron's oscillations is shortened to $\lambda' = \Lambda/\gamma$; so the frequency at which the electron radiates is increased from $\nu = v/\Lambda$ to $\nu' = \gamma v/\Lambda$.

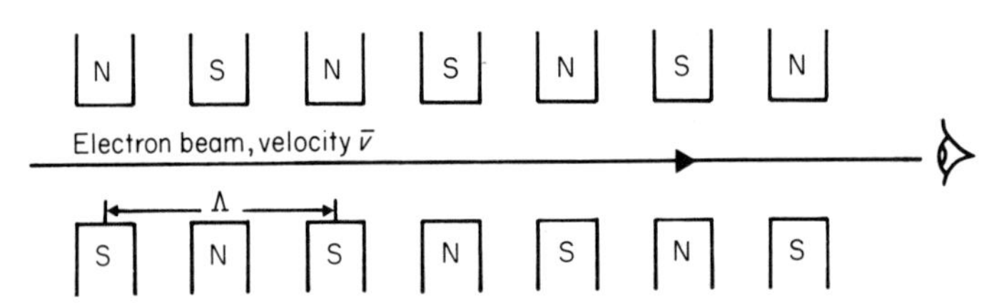

Figure 4.9 The undulator structure: the electron beam traverses an arrangement of magnets whose fields alternate in direction with spatial period Λ.

A different observer in the laboratory reference frame, placed as the eye in figure 4.9, sees the frequency radiated by the electron which is moving towards him Doppler shifted by a further factor of γ, so that altogether the observed frequency is proportional to γ^2. Actually, the factor turns out to be twice this, as can be seen by a more careful analysis.

Undulator radiation may also be regarded as synchroton radiation emitted from a succession of curved orbits arranged in such a way that the light wave can keep in step with the electrons and produce a long, coherent wave train. If electrons can be bunched so as to concentrate them into short packets with a linear size comparable with that of the wavelength, a very large energy can be radiated; and Motz in his 1951 paper[54] calculated that the power which could be emitted coherently in this way is of the order of several tens of kW, in the millimetre band.

Early experiments were done in 1953 by Motz and co-workers[56] at Stanford by using high-energy electrons from a special linear accelerator designed for the injection system of the Mark III Stanford linear accelerator (LINAC). With a 3 MeV electron beam, radiation in a band of wavelength less than 1.9 mm was observed at a power level estimated to be between 10 and 100 W.

Moreover, under the influence of incoming radiation, the beam of electrons can be made to radiate in such a way as to amplify the incoming electromagnetic wave. The incoming wave stimulates the undulating electron in order to amplify it. Classical electronic devices may also often be looked at in this way, and the undulator as an amplifier is closely analogous to the travelling-wave tube amplifier invented by R Kompfner in 1947[57]. In 1959 Motz realised[58] that coherent amplification may be obtained without a slow-wave structure as in the travelling-wave amplifier, but by means of an undulator[59].

Notes

1. See for example C H Townes, *Science* **159**, 699 (1968); Copyright 1968 by the American Association for the Advancement of Science
2. J Weber, *Amplification of microwave radiation by substances not in thermal equilibrium, Trans. IRE Professional Group on Electron Devices* PGED-3, 1 (June 1953)
3. In his paper Weber makes no distinction between levels and states (which is correct provided the degeneracy of states is one)
4. *Masers – Selected Reprints with Editorial Comment* edited by J Weber (Gordon and Breach: New York, 1967) p51
5. C H Townes himself in a work written in collaboration with L E Alsop, J A Giordmaine and T C Wang (*Phys. Rev.* **107**, 1450 (1957)) felt the need to give some historical references. After having recalled the importance of the experiment by E M Purcell and R V Pound (*Phys. Rev.* **81**, 279 (1951)) which we considered in Chapter 3, he remembered the works by Weber and by Basov and Prokhorov and mentioned a statement made by A H Nethercot on his behalf at the *Symposium on Sub-Millimeter Waves* at the University of Illinois in May, 1951
6. Later on Townes together with Schawlow wrote a well known book: C H Townes and A L Schawlow, *Microwave Spectroscopy* (McGraw Hill: New York, 1955).
7. C H Townes in *Laser Focus*, August 1978, p52
8. The inversion spectrum of ammonia was the first and most thoroughly studied microwave spectrum. It was originally observed by C E Cleeton and N H Williams (*Phys. Rev.* **46**, 235 (1934)). Its fine structure was observed by B Bleaney and R P Penrose (*Nature* **157**, 339 (1946)) and independently by W E Good (*Phys. Rev.* **70**, 213 (1946))
9. J P Gordon, *Phys. Rev.* **99**, 1253 (1955)
10. J P Gordon, H J Zeiger and C H Townes, *Phys. Rev.* **99**, 1264 (1955)

11 The Townes' maser theory, already partly developed in the works mentioned in notes 10 and 12, was completed in the paper by K Shimoda, T C Wang and C H Townes (*Phys. Rev.* **102**, 1308 (1956))
12 J P Gordon, H J Zeiger and C H Townes, *Phys. Rev.* **95**, 282 (1954)
13 F Bloch, *Phys. Rev.* **70**, 460 (1946)
14 N Bloembergen, E M Purcell and R V Pound, *Phys. Rev.* **73**, 679 (1948)
15 N G Basov and A M Prokhorov, as we shall see in §4.4, gave a theoretical treatment of the maser in terms of a complex susceptibility. This method was later used by many authors; see for example R W De Grasse, E O Schultz-DuBois and H E D Scovil, *Bell Syst. Tech. J.* **38**, 305 (1959); A M Clogston, *J. Phys. Chem. Solids* **4**, 271 (1957); P W Anderson, *J. Appl. Phys.* **28**, 1049 (1957) and A Javan, *Phys. Rev.* **107**, 1579 (1957)
16 Theoretical studies were performed by K Shimoda, H Takahasi and C H Townes (*J. Phys. Soc. Japan* **12**, 686 (1957)), R V Pound (*Ann. Phys., NY* **1**, 24 (1957)), M W P Strandberg (*Phys. Rev.* **106**, 617 (1957)). Measurements were done by J P Gordon and L D White (*Phys. Rev.* **107**, 1728 (1957)). The role of spontaneous emission was emphasised by R H Dicke at the *Symposium on Amplification by Atomic and Molecular Resonance,* Asbury Park, New Jersey (1 March 1956) and reported by J Weber in *Rev. Mod. Phys.* **31**, 681 (1959)
17 C H Townes, *J. Inst. Elec. Comm. Eng. (Japan)* **36**, 650 (1953) (in Japanese)
18 See for example A Dymanus, *Int. Rev. Sci. Phys. Chem., Ser. 2* **3** (1975); D C Laine, *Adv. Electr. Electron Phys.* **39**, 183 (1975). In this last reference advances in molecular beam masers in general are reported
19 V A Fabrikant, M M Vudynskii and F Butayeva, USSR Patent no. 123209 submitted 18 June 1951 and published in 1959. Supplemented by USSR Patent no. 148441
20 N G Basov and A M Prokhorov, *Dokl. Akad. Nauk* **101**, 47 (1955)
21 N G Basov and A M Prokhorov, *Zh. Eksp. Teor. Fiz.* **27**, 431 (1954)
22 N G Basov and A M Prokhorov, *Zh. Eksp. Teor. Fiz.* **28**, 249 (1955) (in Russian) and *Sov. Phys.-JETP* **1**, 184 (1955) (in English)
23 N G Basov and A M Prokhorov, *Trans. Faraday Soc.* **19**, 96 (1955)
24 N G Basov, *Radioteck. Elektron.* **1**, 752 (1956); see also Basov's doctoral dissertation *Molecular Oscillators* (Lebedev Institute of Physics: Moscow, 1956)
25 N Bloembergen, *Phys. Rev.* **104**, 324 (1956)
26 J Weber in his excellent review paper on Masers (*Rev. Mod. Phys.* **31**, 681 (1959)) refers on p 692 to the fact that Ali Javan was also working on a three-level maser idea.
27 M W P Strandberg as reported by J Weber in *Masers* (Gordon and Breach: New York, 1967) p 3, noted in 1955 that the use of low temperatures and paramagnetic ions would result in improved intrinsic amplification. See also M W Strandberg, *Proc. IRE* **45**, 92 (1956)
28 H E D Scovil, G Feher and H Seidel, *Phys. Rev.* **105**, 762 (1957)
29 A L McWhorter and J W Meyer, *Phys. Rev.* **109**, 312 (1958)
30 G Feher and H E D Scovil, *Phys. Rev.* **105**, 760 (1957)
31 G Makhov, C Kikuchi, J Lambe and R W Terhune, *Phys. Rev.* **109**, 1399 (1958)

32 J E Geusic, *Phys. Rev.* **102**, 1252 (1956)

33 A A Manenkov and A M Prokhorov, *Sov. Phys.-JETP* **1**, 611 (1955); M M Zaripov and I I Shammonin, *Sov. Phys.-JETP* **3**, 171 (1956)

34 J O Artman, N Bloembergen and S Shapiro, *Phys. Rev.* **109**, 1392 (1958)

35 R W De Grasse, E O Schultz-DuBois and H E D Scovil, *Bell Syst. Tech. J.* **38**, 305 (1959)

36 H R Sent, F E Goodwin, J E Kiefer and K W Cowans, *IRE Trans. Mil. Electr.* MIL-5, 58 (1961)

37 J J Cook, L G Cross, M E Bair and R W Therune, *Proc. IRE* **49**, 768 (1961); J V Jelley, *Proc. IEEE* **51**, 30 (1963) gives a list of various maser installations at that time

38 A A Penzias and R W Wilson, *Astrophys. J.* **142**, 419, 1149 (1965). See also W J Tabor and J T Sibilia, *Bell Syst. Tech. J.* **42**, 1863 (1963)

39 H Motz, *J. Electr. Control* **2**, 571 (1957)

40 J Combrisson, A Honig and C H Townes, *C.R. Acad. Sci., Paris* **242**, 2451 (1956)

41 G Feher, J P Gordon, E Buehler, E A Gere and C D Thurmond, *Phys. Rev.* **109**, 221 (1958)

42 P F Chester, P E Wagner and J G Castle, *Phys. Rev.* **110**, 281 (1958)

43 H E D Scovil and E O Schultz-DuBois, *Phys. Rev. Lett.* **2**, 262 (1959); *Prog. Cryog.* **2**, 173 (1961)

44 W H Carter, *Science* **126**, 810 (1957); H H Theissing, P J Caplan, F A Dieter and N Rabbiner, *Phys. Rev. Lett.* **3**, 460 (1959); I Weider, *Phys. Rev. Lett.* **3**, 468 (1959)

45 V E Derr, J J Gallagher, R E Johnson and A P Sheppard (*Phys. Rev.* **5**, 316 (1960)) and N Knable (*Bull. Am. Phys. Soc.* **6**, 68 (1961)) claimed to have obtained stimulated emission but the effects were extremely weak

46 D P Devor, I J D'Haenens and C K Asawa, *Phys. Rev. Lett.* **8**, 432 (1962)

47 H Hsu and F K Tittel, *Proc. IEEE* **51**, 185 (1963)

48 J F Ready and D Chen, *Proc. IRE* **50**, 329 (1962)

49 H Weaver, D R Williams, H Nannilou Dieter and T W Lum, *Nature* **208**, 29 (1965); H Weaver, H Nannilou Dieter and D R Williams *Astrophys. J. Suppl.* **16**, 10, 146 and 219 (1968); M M Litvak, A L McWhorter, M L Meeks and H J Zieger, *Phys. Rev. Lett.* **17**, 821 (1966); F Perkins, T Gold and E E Salpeter, *Astrophys. J.* **145**, 361 (1966); A H Cook, *Nature* **210**, 611 (1966)

50 D M Rank, C H Townes and W J Welch, *Science* **174**, 1083 (1971); A H Cook, *Celestial Masers* (Cambridge University Press: London, 1977)

51 A Javan, *Bull. Am. Phys. Soc., Ser. 2* **3**, 213 (1958)

52 See for example J L Hirshfield and V L Granatstein, *IEEE Trans. Microwave Theory Tech.* **MTT-25**, 522, (1977); J L Hirshfield in *Infrared and Millimeter Waves* edited by K J Button (Academic: New York, 1979) vol 1, p 1; V L Granatstein and P Sprangle, *IEEE Trans. Microwave Theory Tech.* **MTT-25**, 545 (1977); V A Flyagin, A V Gaponov, M I Petelin and V K Yulpatov, *IEEE Trans. Microwave Theory Tech.* **MTT-25**, 514 (1977); K Mizuno and S Ono, *Infrared and Millimeter Waves* edited by K J Button (Academic: New York, 1979) vol 1 p213

53 H Motz, *J. Appl. Phys.* **22**, 527 (1951)

54 In his paper (note 53) in a note added in proof Motz also said that according to reports of the Electronics Research Laboratory at MIT, P D Coleman was working on a scheme similar to the one he (Motz) was suggesting. Prior reference to the problem treated in the paper was also found in a paper by V L Ginsburg, *Radiation of microwaves and their absorption in air, Bull. Acad. Sci. USSR Phys. Ser.* **9**, no. 2, 165 (1947) (in Russian)

55 H Motz, W Thon and R N Whitehurst, *J. Appl. Phys.* **24**, 827 (1953)

56 R Kompfner, *Proc. IRE* **35**, 124 (1947)

57 H Motz and M Nakamura, *Brooklyn Polytechnic Symposium on Millimeter Waves* (1959) p155

58 Motz's theory in note 53 has recently been further developed by N M Kroll, *Novel Sources of Coherent Radiation* (Addison Wesley: New York, 1978) p115

59 See for example R Q Twiss, *Aust. J. Phys.* **11**, 564 (1958)

5

The Laser

5.1 Introduction

Once the maser had been invented, the idea went around that what was then referred to as an *optical maser* – a light generator based on stimulated emission – could be similarly constructed. In both the USA and the Soviet Union researchers were working independently on the problem. Amongst those in the forefront of this work were, in the USSR, V A Fabrikant and, in the USA, Robert H Dicke (1916-). Fabrikant, as we saw in Chapter 4, had first made his proposals back in the 1940s[1]. In 1954 Dicke, as we shall see in §5.5, published a paper introducing the concept of *superradiance*[2], which laid the foundation for his subsequent discussion of the 'optical bomb'[3]. Dicke, in 1956, suggested the use of the Fabry–Perot interferometer as the resonant cavity and, in 1958, patented this proposal[4].

It was not until later that concentrated efforts towards using stimulated processes to obtain visible radiation were made. In the Soviet Union this research was mainly performed at the Lebedev Institute under Basov and Prokhorov, while in the United States the main workers were Gordon Gould (1920-) – whose story will be told in §5.4 – and Charles H Townes and Arthur L Schawlow. The latter were the first to publish detailed and exhaustive proposals which subsequently led to the construction of various types of laser; and so, although the first lasers to become operational were not the kind which had been considered by them, we shall begin by describing their proposals.

5.2 The Townes and Schawlow proposal

In 1957 Charles H Townes began to consider the problems connected with making maser-type devices work at optical wavelengths. Townes carried out his work in close collaboration with Arthur L Schawlow, then a research physicist at Bell Laboratories.

Arthur L Schawlow was born in Mount Vernon, New York in 1921 and obtained his BA, MA and PhD in physics in Toronto, Canada. Soon after the war he wrote to I I Rabi at Columbia University, who suggested that he applied for a post-doctoral fellowship to work under Townes. This fellowship was given by Carbide and Carbon Chemical Corporation, a division of Union Carbide, for research on the applications of microwave spectroscopy to organic chemistry. Schawlow and Townes became friends, often dining together at the Columbia Faculty Club where there was a table reserved for a group of physics and mathematics professors. Schawlow later married Townes' sister.

In 1951 Schawlow accepted a post working in solid-state physics at the Bell Laboratories, where he became interested in superconductivity and nuclear resonance. At this time he was also writing a book with Townes on microwave spectroscopy[5] and spent nearly every Saturday at Columbia.

In 1957 Schawlow began to think about the possibility of building some kind of infrared maser and writes[6]:

> A few weeks later, about October 1957, Charles Townes visited Bell Labs. and we had lunch. Townes had been consulting with the Laboratories for about a year, but his contacts were with the maser people and I had not had any serious discussion with him. He told me then that he was interested in trying to see whether an infrared or optical maser could be constructed, and he thought it might be possible to jump over the far infrared region and go to the near infrared or perhaps even the visible portion of the spectrum. He had some notes and said that he would give me a copy. We agreed that it might be worthwhile for us to collaborate on this study and so we began.

This story is confirmed also by Townes who writes[7]:

> I discovered that my friend Arthur Schawlow, then at the Bell Telephone Laboratories, had also been thinking along somewhat similar lines, and so we immediately pooled our thoughts. It was he who initiated our consideration of a Fabry–Perot resonator for selection of modes of the very short electromagnetic waves in the optical region.
>
> This very likely had something to do with the fact that Schawlow had first been trained as a spectroscopist and had done his thesis with a Fabry–Perot . . .

So, Townes decided to give his notes to Schawlow. In these notes calculations had been made using thallium atoms, which, it was intended, should be excited from the ground state (6p) to a higher one (6d or 8s) through the ultraviolet light of a thallium lamp. Schawlow writes[6]:

Such lamps were in use in Kusch's laboratory at Columbia University for experiments on optical excitation of thallium atoms in an atomic beam resonance experiment.

Townes had discussed with Gordon Gould, a student of Kusch's who was working on the atomic beam experiment, the properties of thallium lamps to find out how much power could be expected from them.

On 14 September 1957 Townes, asked a Columbia graduate student, Joseph Antony Giordmaine, to sign a notebook wherein a light resonator was described: this consisted of a glass box with four mirror walls, and used a thallium lamp to energise thallium inside the cavity. Schawlow quickly demonstrated that this thallium scheme of Townes would not have worked easily and so they started to search for other materials. In the meantime Townes calculated the number of excited atoms needed (the equations later being published[8]) and performed some experiments. They then turned their attention to the cavity to be used.

As Townes' report[7] mentioned, Schawlow made an important contribution to the final choice (a Fabry–Perot type cavity). Schawlow himself recalled[6] that when he had been a student in Toronto he had become familiar with the use of the Fabry–Perot interferometer during some research he carried out on the hyperfine structure in atomic spectra under the direction of Professor Malcolm F Crawford:

> I had in mind from the beginning something like the Fabry–Perot interferometer I had used in my thesis studies. I realized, without even having looked very carefully at the theory of this interferometer, that it was a sort of resonator in that it would transmit some wavelengths and reject others.

Later on he wrote some of his deliberations in a notebook. On 29 January 1958 he asked Solomon L Miller, a graduate student of Townes' working at Bell Laboratories (who later went to IBM) to sign this notebook. Somehow Schawlow had understood that a good resonator, which reduced the large number of modes existing in a cavity and therefore prevented the hopping from one mode to another during operation, was simply one made by two parallel plane mirrors some distance apart, without any other reflecting part.

During the spring Schawlow and Townes decided to write up this work for publication. It was customary at the Bell Laboratories to circulate manuscripts among colleagues prior to publication in order to obtain technical comments and improvements. A copy was also sent to Bell's patent office to see whether it contained any invention worth patenting.

As a result of this process colleagues asked them to write more on the modes but the patent office at first refused to patent either their amplifier or their optical frequency oscillator because[7] 'optical waves had never been of any importance to communications and hence the invention had little bearing on Bell System interests'. However, upon Townes' insistence, a patent request was filed and it was delivered in March 1960[9]. The paper itself was received on 26 August 1958 and published in December of the same year in *Physical Review*[8]. Its authors were later awarded the Nobel Prize; C H Townes in 1964, as we have seen, for his invention of the maser and proposal for the laser, and A L Schawlow in 1981 for a related subject: laser spectroscopy.

5.3 Townes' and Schawlow's ideas

In their paper entitled *Infrared and Optical Masers* Schawlow and Townes observed that, although it was possible in principle to extend the maser techniques into the infrared and optical region to generate very monochromatic and coherent radiation, a number of new aspects and problems arise which require both a quantitative re-orientation of the theoretical discussion and a considerable modification of the experimental techniques used.

The declared purpose of the paper was to discuss the theoretical aspects of maser-like devices for visible or infrared wavelengths and to outline the design considerations, so as to promote the realisation of this new kind of maser, called by them an optical maser (later to be called a *laser*, where 'L' stood for *light*). The principal points considered were the choice of a cavity and its mode selection properties; the expression of the gain of the device; and some proposals on active materials.

The most immediate problem was the realisation of a resonant cavity. In the case of the maser an ordinary microwave cavity with metallic walls had been used. By suitable design of such a cavity, it was possible to obtain just one resonant mode oscillating near the frequency corresponding to the radiative transitions of the active system.

In order to obtain such a single, isolated mode, the linear dimension of a cavity needs to be of the order of one wavelength which, at infrared frequencies, would be too small to be practical. Hence, it is necessary to consider cavities of dimensions large compared with a wavelength and which can therefore support a large number of modes within the frequency range of interest. Townes and Schawlow realised then that it was necessary to increase selectively the Q of only certain modes. After some general considerations, the authors' choice was a Fabry-Perot interferometer constructed by two perfectly reflecting plane parallel end walls. In it, a plane wave travels back and forth between the two mirrors and if the relation $L = l(\lambda/2n)$ is fulfilled (where λ is the vacuum wavelength, n is the refractive index, l some integer number and L the distance between mirrors) stationary waves are created. (The paper deals in depth with the choice of the cavity and its mode selection properties – mode hopping, mode instability due to changes in cavity dimensions, and mode competition were clearly identified phenomena.)

Having, it was hoped, generated by this process a light signal, it would then have to be collected. It was suggested that one of the mirrors be partially reflecting, so as to allow a beam of light incident on it to be partially transmitted outside the cavity. The idea of adopting a cavity of dimensions much larger than the wavelength had been suggested not only by the practical impossibility of constructing one of the correct dimension for resonance, but also by the fact that the resonator should contain a reasonable quantity of active material. (As discussed later, (§ 5.5) other authors had also proposed the use of such a cavity.)

Another problem treated in the paper was the determination of the minimum number of molecules or atoms of the active material which should be at the higher energy level to allow the generation of light by stimulated emission. This problem was simplified by considering a material with only two energy levels, E_1 and E_2, with $E_2 > E_1$, and with a population N_1 and N_2 respectively.

The condition for oscillation was obtained by requiring that the power produced by stimulated emission be as great as that lost on the cavity walls or due to other types of absorption. That is

$$(\mu' E/\hbar)^2 (h\nu N/4\pi\Delta\nu) \geqslant (\overline{E^2}/8\pi)(V/t), \tag{5.1}$$

where μ' is the matrix element for the emissive transition, E^2 is the mean square of the electric field, $N = N_2 - N_1$, V is the volume of the cavity, t is the time constant for the rate of decay of the energy and $\Delta\nu$ is the half-width of the resonance at half maximum intensity if a Lorentzian shape is assumed. The decay time was simply calculated considering a plane wave reflected back and forth many times between the two mirrors of the cavity at a distance D from each other. The wave undergoes reflection every time it travels a distance D, and the rate of loss of energy W is given by the equation

$$\mathrm{d}W/\mathrm{d}t = -c(1-\alpha)\, W/D, \tag{5.2}$$

where α is the reflection coefficient of the cavity wall. The solution of equation (5.2) is

$$W = W_0 \exp\left[-c(1-\alpha)\, t/D\right],$$

from which the decay time t results,

$$t = D/c\,(1-\alpha).$$

The condition for oscillation may be conveniently related to the lifetime τ of the state due to spontaneous emission of radiation by a transition between the two levels in question. This lifetime is given by well known theory as

$$\tau = hc^3/(64\pi^2\nu^3\mu'^2).$$

Now, the rate of stimulated emission due to a single quantum in a single mode is just equal to the rate of spontaneous emission into the same single mode (cf equation (2.6)).

Hence $1/\tau$ is this rate multiplied by the number of modes p which are effective in producing spontaneous emission. Assuming a single quantum present in a mode at the resonant frequency, the condition for instability can then be written

$$nh\nu/p\tau \geqslant h\nu/t,$$

or

$$n \geqslant p\tau/t,$$

which is equivalent to expression (5.1).

The minimum power which must be supplied in order to maintain n systems in excited states is finally derived as

$$P = nh\nu/\tau = ph\nu/t.$$

The number of effective modes in this expression is

$$p = \int p(\nu) f(\nu) \, \mathrm{d}\nu,$$

where $f(\nu)$ is the line profile, and $p(\nu)\,\mathrm{d}\nu$ is the number of modes between ν and $\nu + \mathrm{d}\nu$, which is

$$p(\nu)\,\mathrm{d}\nu = 8\pi\nu^2 V\,\mathrm{d}\nu/c^3.$$

For a Lorentzian lineshape it is

$$p = 8\pi^2\nu^2 V\Delta\nu/c^3.$$

For a line broadened by Doppler effects it is

$$p = 8\pi^2\nu^2 V\Delta\nu/(\pi \ln 2)^{1/2}\,c^3.$$

The minimum power is therefore, e.g. for a Lorentzian line,

$$P = 8\pi^2 h\nu^3 V\Delta\nu/c^3 t,$$

which shows an important property — the scaling of pumping power with ν^3.

Monochromaticity of a maser oscillator was also considered and Schawlow and Townes observed that this property is very closely connected with the noise properties of the device as an amplifier.

The noise present was conceptually analogous to the one present in the radio-wave range, but of a different nature. In an electronic oscillator, noise is essentially thermal and is spread over the whole frequency range of the useful band of the oscillator. In the laser, noise consists of the spontaneous emission of radiation by the active material. Schawlow and Townes, by modifying the calculation previously developed for masers, discussed in their paper the interaction between an atomic line and one mode of the cavity, whose frequency coincided approximately with the centre of that atomic line.

They obtained in this way an analytic expression for the linewidth and found it was of the order of one millionth of the linewidth corresponding to the cavity mode. This gave an estimate of the spectral purity and demonstrated the excellent monochromaticity of the laser radiation. They obtained as a theoretical lower limit for the noise linewidth

$$\delta\nu = 4\pi h\nu(\Delta\nu)^2/P,$$

which is analogous to that of the maser, where the thermal noise, kT, is replaced by the noise resulting from the stimulated emission of radiation. It is interesting to observe that they did not note that in the optical case the quantity $\Delta\nu$ refers to the half-width of the resonant mode and not to the linewidth of the spectral line, which is generally much larger.

A paragraph in the paper was then devoted to discussing a specific example using optical pumping. As an example of a particular system for an infrared maser they considered atomic potassium vapour pumped at 4047 Å. Schawlow had even performed some preliminary experiments on commercial potassium lamps and had asked Robert J Collins, a spectroscopist at Bell Laboratories, to measure the power output of these lamps. Another possibility was caesium, which they proposed could be excited with a helium line.

They also considered solid-state devices, although they were not very optimistic. In this case they wrote:

> The problem of populating the upper state does not have as obvious a solution in the solid case as in the gas. Lamps do not exist which give just the right radiation for pumping. However, there may be even more elegant solutions. Thus it may be feasible to pump to a state above one which is metastable. Atoms will then decay to the metastable state (possibly by non-radiative processes involving the crystal lattice) and accumulate until there are enough for maser action. This kind of accumulation is most likely to occur when there is a substantial empty gap below the excited level.

The work by Schawlow and Townes created considerable interest and many laboratories started to search for possible materials and methods for optical masers.

Townes, with his group at Columbia, started efforts to build an optical maser with potassium. He worked with two graduate students, Herman Z Cummins and Isaac Abella.

At the same time Oliver S Heavens, now Professor of Physics at York University, York, England, who was then already a world expert on highly reflecting mirrors, joined this group. Townes had in fact realised that cavity mirrors were the most delicate point of the system. The system they were studying had internal mirrors and we may today ascribe their operational failure as being due to the degradation of the mirror coatings owing to bombardment from the ions of the electrical discharge in the tube. A report on their work was published in 1961[10]. The caesium laser, which was also considered in the paper, was later realised by S Jacobs, G Gould and P Rabinowitz in 1961[11].

Schawlow, at Bell, had begun to consider ruby as a possible solid-state material. Others, too, were considering ruby: in Japan, Saturo Sugano and Y Tanabe; Irvin Weider at Westinghouse Research Laboratories; and Stanley Geschwind at Bell Laboratories.

In 1959 Schawlow[12] also suggested using ruby, but observed that the R lines were not useful for laser action:

> There is a broad absorption band in the green and others in the ultraviolet. When excited through these bands, the crystal emits a number of sharp bands in the deep red (near 7000 Å). The two strongest lines (at 6919 Å and 6934 Å) go to the ground state, and are not suitable for laser action. However, the strongest satellite line (at 7009 Å) ... goes to a lower state which is normally empty at liquid-helium temperatures, and might be usable ... The structure of a solid-state maser could be especially simple. In essence, it would be just a rod with one end totally reflecting and the other end nearly so. The sides would be left clear to admit pumping radiation.

In the same work Schawlow quoted the studies by Ali Javan at Bell Laboratories on the energy transfer in collisions between two kinds of gases in a mixture, which could give rise to the necessary population inversion.

5.4 The Gordon Gould story

Schawlow and Townes were not the only ones who foresaw the potential of an optical maser and who had faith that methods would soon be found for the creation of a medium with a negative absorption coefficient in the optical range.

Gordon Gould, who after an MSc at Yale in 1943 had been a graduate student at the Radiation Laboratory of Columbia University where Townes taught, filed a patent application in the United States on 6 April 1959, as evidenced by a series of British patents[13] based on his US patent application. These disclose material similar to that in the Schawlow-Townes patent. Although Gould did not publish his findings in the customary manner, his patent disclosures are interesting because they formed the basis of a legal contest concerning the invention of lasers.

It is interesting to obtain some insight into this story. In October 1957, according to the Gordon Gould deposition at a trial (TRG Inc. versus Bell Laboratories, for the priority of laser invention) he was developing the possibility of using the Fabry-Perot arrangement for use as a laser resonator, when he received, at his home, a telephone call from Townes. Townes, who had his office near Gould on the tenth floor of the Physics Building at Columbia, wanted some information on the very bright thallium lamps that Gould had used when he was research assistant at Columbia Radiation Laboratory[14]. This telephone call is registered in Townes' notebook. This call made Gould very excited and led him to rush to finish his studies as quickly as he could. On Friday 16 November 1957, Gould and his wife, Dr Ruth Frances Hill Gould, Assistant Professor of Radiology at Columbia and graduate in Physics at Yale, went to the proprietor of a candy store (who was a public notary and a friend of Gould's wife and of his family) who there and then put his seal on the first nine pages of Gould's laboratory notebook which contained the work *Some rough calculation on the feasibility of laser light amplification by stimulated emission of radiation.*

By an irony of fate it was Townes who introduced Gould to the use of a signed notebook as a method of establishing priority of claim to an invention! And it is on this notebook that TRG, the firm where Gould worked after leaving Columbia, based their action against Bell Laboratories for the prior claim to the invention of the laser. The trial was held on 8 December 1965 in Washington DC at the US Court of Customs of Patent Appeal.

Gould's obsession about the laser had already cost him dearly. His professor of physics, Professor Polykarp Kusch, had not approved a thesis on this argument and, in March 1958, Gould had left Columbia for TRG Inc. without obtaining his PhD.

At the trial Dr Alan Berman, a physicist friend and bridge partner of Gould, who was Associate Director at Hudson Laboratories, Dobbs Ferry, Columbia, testified to a conversation he had with Gould in August 1958 at a beach party at Fire Island, New York (Gould at that time was with TRG). Berman said:

> I again noted that Gordon was working on the laser to the detriment of his thesis. I was annoyed at him...he did not make any attempt to publish it in properly defined channels, and I thought this was an unscientific thing to do.

It is worth noting that if Gould had followed Berman's advice he could have sent *Physical Review* a paper that could have been published simultaneously with the one by Schawlow and Townes!

In September 1958, Lawrence Goldmuntz, president of TRG, became aware that Gould was wasting his time working on a private project. Gould and Goldmuntz discussed the research on the laser and TRG took on the Gould project as its own. On 16 December 1958, Goldmuntz asked for $200 000 from the Aerojet-General Corp. of El Monte, California, that at that time had 18% of TRG. $300 000 was also requested from the Advanced Research Projects Agency (ARPA) of the Pentagon to enable the use of Gould's work on the laser for optical radar, range finding and communication systems.

By coincidence Gould and Townes, at almost the same time, each had in their hands the work of the other. Gould received a preprint of the Schawlow and Townes paper from Dr Maurice Newstein, a researcher at TRG: Townes, as an adviser to the federal government, was reading the 200 page proposal submitted to ARPA. Other people, too, saw the latter and gave a favourable reply to it: TRG received $998 000 for the project.

Unfortunately, Gould was unable to work on his project at TRG because it was classified and he had not got the necessary security clearance. However, he was the first to develop the cw Cs optically pumped laser[11].

The Gould story does not end here. Today (1982) he is Vice President of Optelecom. Inc. of Gaithersburg, Maryland, a firm that develops communication systems with optical fibres for military purposes. On 11 October 1977, after years of efforts he finally received his patent for an optically pumped laser amplifier (no. 4.053.845).

The request for this patent was filed in 1959 when he was still working at TRG Inc. The patent was finally issued after 18 years of waiting; a record, perhaps, if one considers that three or four years is the usual waiting time! (In the meantime he had had to change direction, and, as he had not been able to get a laser generator patent, he had asked for and obtained an amplifier patent instead, based on the same disclosures.) This long delay meant that the Gould patent came into operation just when those of Schawlow and Townes, issued in 1960, ended. (In the US patents run for 17 years.)

Gould then gave the management of his long-awaited patent to Refac Technology Development Corp. of New York, which started a campaign to demand considerable royalties (from 3.5 to 5%) from laser manufacturing companies. These, of course, had no intention of paying any money to Gould.

Gould, who considered the issue of the patent to be a victory over his 'enemies', declared:

> I really felt vindicated. I'm beginning to believe that it's all come true. But emotionally, after 18 years, it's a little hard to accept.

But the manufacturing companies, in turn, had started a campaign against the royalties requested, which would, if payed, cost them in one year as much as was paid to Townes in 17 years – about one million dollars. Even now the whole story has not been told, but Gould is receiving royalties. A second patent staking out three broad claims on laser applications was issued on 17 July 1979 to Gordon Gould by the United States Patent Office. It was entitled *Method of and apparatus for heating employing laser amplifiers* and received the no. 4.161.436. It was the second to result from the application Gould filed in 1959.

The applications patent covers broad subjects, including laser photochemistry, fusion isotope separation and materials working. The patent contains three claims, each prefaced by a description of the operation of a laser amplifier. The claims described three ways of 'energising' a material:

(*a*) irradiation of a 'material which includes a mixture of reactable chemicals whereby said emitted substantially collimated light (from the laser) promotes reaction between said chemicals';

(*b*) irradiation of a material 'so that at least a portion of said material dissociates in response to said emitted substantially collimated light';

(*c*) irradiation of a material in a process 'in which said material responds to said emitted substantially collimated light by undergoing an exothermic process thereby releasing additional energy.

The second patent may not be Gould's last. Two more patents are pending, one covering gas-discharge lasers and the other Brewster-angle windows. The Patent Office is still examining additional claims. Gould also received a Canadian patent for discharge-pumped lasers in 1972 (no. 907.110).

5.5 The Dicke coherence-brightened laser

Among those who have the right to be remembered as having paved the way for the invention of lasers, R H Dicke, as we saw earlier (§ 5.1), deserves a particular mention. He not only developed novel and original ideas for the production of coherent radiation in the microwave and IR spectrum, without making recourse to the feedback concept but also was among the first to suggest the use of a Fabry–Perot cavity. These two points are different and must be treated separately.

As a proponent of new coherent sources, Dicke, in the paper *Coherence in Spontaneous Radiation Processes*, published in the January 1954 issue of *Physical Review*[2], treated for the first time spontaneous emission of electromagnetic radiation from an atomic (or molecular) system in a correlated state. In this paper Dicke developed the concept that molecules do not radiate independently of each other. This latter, simplified, picture overlooks the fact that all the molecules are interacting with a common radiation field and hence cannot be treated as being independent. By considering a radiating gas as a single quantum mechanical system, energy levels corresponding to certain correlations between individual molecules were described. Using a point-source model, he defined collective energy eigenstates of two-level atoms.

The quantum numbers labelling these Dicke states were the energy quantum number M and the cooperation number J. If N is the total number of atoms in the sample, then $0 \leqslant J \leqslant N/2$ and $-J \leqslant M \leqslant J$. Dicke found that the initial radiation rate from a system in such a collective state is proportional to $(J+M)(J-M+1)$.

A state with $J \simeq N/2$ and $M \simeq 0$ has the maximum emission rate, proportional to N^2. Such a state he called *superradiant*.

Two ways in which a superradiant state may be excited were described. The first is if all the molecules are excited; the second is to have the gas in its ground state and irradiate it with a pulse of radiation. This latter excitation method was clearly related to Hahn's photon echo method, and Dicke made several references in his paper to magnetic resonance experiments.

A laser which does not employ mirrors in order to produce feedback amplification, but rather is a source of spontaneous emission of radiation with the emission process taking place coherently, was first discussed briefly by Dicke in a talk before the American Physical Society on 23 January 1953. It was also discussed at the Fourth Congress of the International Commission on Optics held in Boston in 1956, and published in the Journal of the Optical Society of America[15].

In 1953, the words 'maser' and 'laser' having not yet been coined, Dicke called his device an *optical bomb* because he predicted that a laser of this type would be characterised by an unusually short and intense light burst[3].

A sharp, intense source was also described in principle by Dicke in 1957 in a paper written in collaboration with R B Griffiths[16]. It was based on the use of resonance radiation coherently scattered from a collimated beam of atoms. Dicke observed that if the excitation were carried out with monochromatic radiation the coherent reradiation is simply the coherent forward scattering familiar as a phenomenon of resonance fluorescence.

In the optical region a rather simple technique for inducing coherent radiation is therefore to excite coherent forward scattering using resonance fluorescence. He also proposed the separation of the coherently scattered light from the exciting radiation either by placing the atomic beam in one arm of a Mach-Zehnder interferometer or by diffraction from several parallel atomic beam pencils. The latter technique was tested experimentally by observing hyperfine structure in light coherently scattered from a beam of sodium atoms. Apparently this was the first experiment in forward scattering spectroscopy[17].

Superradiance was observed much later by the MIT group of Feld and co-workers in 1973[18]. It received a large amount of theoretical attention[19] and has begun, in recent years, to be studied experimentally[20]. Before the Feld work, Abella *et al*[21], in a paper on photon echoes published in *Physical Review* in 1966, described an experiment in which a dilute ruby crystal was found to emit spontaneously a short pulse of light, the photon echo, at a time τ_s after irradiation by two successive ruby-laser pulses separated by τ_s. In the very beginning of the paper they say:

> The purpose of the first pulse is to excite a superradiant state† exhibiting an oscillating macroscopic electric dipole moment. This dipole moment quickly dephases because of inhomogeneous crystal-field strains, and the atoms then radiate at the normal spontaneous emission rate. The second excitation pulse reverses the dephasing process so that the system rephases at the same rate at which it dephased. When the rephasing process is complete, the macroscopic electric dipole moment is momentarily reformed, and the crystal emits an intense burst of light, the photon echo.

The second important contribution of R H Dicke comes from the patent *Molecular amplification and generation systems and methods* filed in 1956 and issued in 1958[4]. In this patent the use of the Fabry–Perot interferometer is described.

5.6 Soviet research

Fabrikant and his students[22] have a patent bearing the application date 18 June 1951, but which was only published in 1959. At first, the Soviet Patent Office did not accept it. Nor did Fabrikant construct any laser or maser. However, we may recall that in the Soviet Union the formulation of physical laws or the descriptions of physical properties of phenomena can be patented as new inventions. Fabrikant's patent is entitled *A method for the amplification of electromagnetic radiation (ultraviolet, visible, infrared and radio waves), distinguished by the fact that the amplified radiation is passed through a medium which, by means of auxiliary radiation or by other means, generates excess concentration, in comparison with the equilibrium concentration of atoms, other particles, or systems at upper energy levels corresponding to excited states.* Although the purpose of the patent appears to be very general, it

† R H Dicke, *Phys. Rev.* **93**, 99 (1954)

seems that Fabrikant was mainly interested in light emission in a gaseous medium.

Fabrikant foresaw the practical significance of a light amplifier which can be modulated and he recognised the usefulness of enclosing the amplifying material in a resonant structure. Apparently he had been considering the problem even in his early years: according to Lengyel[23], he was already working on it in his doctoral dissertation in 1940! He wrote there:

> For molecular (atomic) amplification it is necessary that N_2/N_1 be greater than g_2/g_1. Such a situation has not yet been observed in a discharge even though such a ratio of populations is in principle attainable ...Under such conditions we would obtain a radiation output greater than the incident radiation and we could then speak of a direct experimental demonstration of the existence of negative absorption.

After filing the patent application more than 10 years later, in 1951, he and his students continued the work, but without any success. They experimented with, among other materials, caesium optically pumped with the 3889 Å He line, but they looked at the wrong transition[24]. Lengyel wrote[23]:

> The work of Fabrikant is historically interesting because he attempted to develop the laser without passing through the maser phase, as his American colleagues did. He arrived relatively early at the essential conditions for obtaining population inversion by optical excitation.
>
> The publication of his work was delayed, however, and it did not exert an appreciable influence on the main development of quantum electronics. This art began as a branch of the microwave technology and not as an extension of optical spectroscopy.

In their work at the Optical Institute of the Academy of Sciences, Fabrikant and Butayeva[24] claimed to have obtained light amplification in a discharge of mercury vapour in which the transfer of excitation produced population inversion and negative absorption. However, Sanders *et al*[25], who had a translation of the work by Lengyel, were not able to repeat their work.

Later on Basov and Prokhorov at the Lebedev Institute in Moscow started research. Basov started his work on lasers in 1957 with research into physical methods of obtaining non-equilibrium states in semiconductors, as will be discussed later. The year after, A M Prokhorov[26] in a letter to the editor of the *Soviet Journal of Experimental and Theoretical Physics* in June 1958, pointed out the possibility of assembling a molecular generator and amplifier (MAG) for waves of wavelength shorter than a millimetre, i.e. in the optical region, by using the rotational transitions of ammonia molecules.

The required population inversion was obtained by letting the molecular beam pass through a suitable quadrupolar separator, as in a maser; the proposed pumping was, substantially, non-optical.

In his letter, Prokhorov proposed a theoretical example of how to realise the MAG, giving information regarding the negative absorption coefficient of the material for which amplification was obtained, and considering a cavity with mirrors for which calculation was made of the Q factor as a function of wavelength.

In 1959 N G Basov[27] proposed a method for producing light by exciting a homogeneous semiconductor with electrical pulses. In a semiconductor in a sufficiently high strength of the electric field, due either to ionisation or to tunnelling effects the concentration of non-equilibrium carriers distributed in a large energy band is strongly increased.

To obtain a population inversion and produce light it is necessary to cut off the applied field rapidly, in a time much shorter than the lifetime of the non-equilibrium carriers. This situation is realised when Gunn domains propagate inside some semiconductors (see Chapter 6). In 1974 Basov obtained the effect in a somewhat different set-up[28].

Notes

1 V A Fabrikant, *Doctoral Dissertation*, FIAN P N Lebedev Physical Institute, Academy of Sciences, USSR (1939). *Transactions of the All-Union Order of Lenin Electrotechnical Institute* **41**, *Electron and Ion devices* pp 236, 254 (1940)

2 R H Dicke, *Phys. Rev.* **93**, 99 (1954)

3 R H Dicke, *The Coherence Brightened Laser* in *Quantum Electronics*, edited by P Grivet and N Bloembergen (Dunod: Paris, 1964) p 35

4 R H Dicke, Patent *Molecular amplification and generation systems and methods* requested in 1956 and obtained in 1958, US Patent 2.851.652 (9 September 1958)

5 The book was published later under the title: C H Townes and A L Schawlow, *Microwave Spectroscopy* (McGraw Hill: New York, 1955)

6 A L Schawlow, *From Maser to Laser* in *Impact of Basic Research on Technology* edited by B Kursunoglu and A Perlmutter (Plenum: New York, 1973) p113

7 C H Townes, *Science* **159**, 699 (16 February 1968)

8 A L Schawlow and C H Townes, *Phys. Rev.* **112**, 1940 (1958)

9 A L Schawlow and C H Townes, US Patent No 2.929.922. The filing date was 30 July 1958. The title *A Medium in which a Condition of Population Inversion Exists*. The patent was issued 22 March, 1960

10 H Z Cummins, I D Abella, O S Heavens, N Knable and C H Townes, in *Advances in Quantum Electronics*, edited by J Singer (Columbia University Press: New York, 1961) p 12

11 S Jacobs, G Gould and P Rabinowitz, *Phys. Rev. Lett.* **7**, 415 (1961) and P Rabinowitz, S Jacobs and G Gould, *Appl. Opt.* **1**, 513 (1962)

12 A L Shawlow, *First Int. Quantum Electronics Conf., September 1959, Columbia*, (Columbia University Press: New York, 1960) p553

13 The contents of Gould's application of 1959 were eventually published by the Patent Office (No. 3.388.314) 11 June 1968. The claims allowed in this patent regards 'apparatus for generating radiation of frequencies higher than those of light' and are worthless, because the instrument constructed on these principles will not produce x-rays. However, in this patent are published Gould's ideas. The British patents are G Gould, Brit. Patent Specs. 953721–953727, published 2 April 1964

14 G Gould had used a thallium lamp to produce optically excited states in beams of thallium, G Gould, *Phys. Rev.* **101**, 1828 (1956)

15 R H Dicke, *J. Opt. Soc. Am.* **47**, 527 (1957)

16 R B Griffiths and R H Dicke, *Rev. Sci. Instrum.* **28**, 646 (1957)

17 W Gawlik and G W Series, in *Laser Spectroscopy* vol 4, edited by H Walther and K W Rothe (Springer: Berlin, 1979) p210

18 N Skribanowitz, I P Herman, J C MacGillivray and M S Feld, *Phys. Rev. Lett.* **30**, 309 (1973)

19 Although not exhaustive, a list of works can be found in: F Haake in *Laser Spectroscopy* vol 4, edited by H Walther and K W Rothe (Springer: Berlin, 1979) p 451; R Glauber and F Haake, *Phys. Lett.* **68A**, 29 (1978); F Haake, H King, G Schröder, J Haus, R Glauber and F Hopf, *Phys. Rev Lett.* **42**, 1740 (1979); F Haake, H King, G Schröder, J Haus and R Glauber, *Phys. Rev.* A **20**, 2047 (1979); R Bonifacio, P. Schwendimann and F Haake, *Phys. Rev.* A **4**, 302, 854 (1971); R Bonifacio and L A Lugiato, *Phys. Rev.* A **11**, 1507 (1975); — *Phys. Rev.* A **12**, 587 (1975); M F H Schuurmans and D Bolder in *Laser Spectroscopy* vol 4, edited by H Walther and R W Rothe (Springer: Berlin, 1979) p 459; G Banfi and R Bonifacio, *Phys. Rev. Lett.* **33**, 1259 (1974); — *Phys. Rev.* A **22**, 2068 (1975); J C MacGillivray and M S Feld, *Phys. Rev.* A **14**, 1169 (1976)

20 D Polder, M F H Schuurmans and Q H F Vrehen, *J. Opt. Soc. Am.* **68**, 699 (1978); — *Phys. Rev.* A **19**, 1192 (1979); H M Gibbs, Q H F Vrehen and H M J Hikspoors, *Phys. Rev. Lett.* **39**, 547 (1977); Q H F Vrehen and M F H Schuurmans, *Phys. Rev. Lett.* **42**, 224 (1979); M Gross, C Fabre, P Pillet and S Haroche, *Phys. Rev. Lett.* **36**, 1035 (1976); A Flusberg, T Mossberg and S R Hartmann, *Phys. Lett.* **58A**, 373 (1976); J Okada, K Ikeda and M Matsuoka, *Opt. Commun.* **27**, 321 (1978); Q H F Vrehen, H M J Hikspoors and H M Gibbs, *Phys. Rev. Lett.* **38**, 764 (1977); Q H F Vrehen in *Laser Spectroscopy* vol 4, edited by H Walther and K W Rothe (Springer: Berlin, 1979) p 471; A Crubellier, C Brechignac, P Cahuzac, P Pillet, ibid p 480; E Campani, G Degan, E Polacco and G Gorini, *Lett. Nuovo Cim.* **26**, 269 (1979); G Alessandretti, F Chiarini, G Gorini and F Petrucci, *Opt. Commun.* **20**, 289 (1977)

21 I D Abella, S R Hartmann and N A Kurnit, *Phys. Rev.* **141**, 391 (1966)

22 V A Fabrikant, M M Vudynskii and F Butayeva, USSR Patent No 123209 submitted 18 June 1951 and published in 1959. Supplemented by USSR Patent no. 148441

23 B A Lengyel, *Am. J. Phys.* **34**, 903 (1966)

24 F A Butayeva and V A Fabrikant, *Investigations in Experimental and Theoretical Physics* (A memorial to G S Landsberg) (USSR Acad. Sci. Publ: Moscow, 1959) pp62–70

25 J H Sanders, M J Taylor and C E Webb, *Nature* **113**, 767 (1962); see also a footnote in W R Bennett Jr, *Appl. Opt. Suppl.* **1**, 24 (1962) and S Jacobs, G Gould and P Rabinowitz, *Phys. Rev. Lett.* **7**, 415 (1961)

26 A M Prokhorov, *Zh. Eksp. Teor. Fiz.* **34**, 1658 (1958); *Sov. Phys.-JETP* **7**, 1140 (1958)

27 N G Basov, B M Vul and Yu M Popov, *Zh. Eksp. Teor. Fiz.* **37**, 587 (1959); *Sov. Phys.-JETP* **10**, 416 (1959); V V Bagaev in *Quantum Electronics* edited by P Grivet and N Bloembergen (Dunod: Paris, 1964) p1899

28 N G Basov, A G Molchanov, A S Nasibov, A Z Obidin, A N Pechenov and Yu M Popov, *Sov. Phys.-JETP Lett.* **19**, 336 (1974)

6

The Laser: Further progress

6.1 Introduction

Immediately after the publication of the paper by Schawlow and Townes (see Chapter 5) a number of people started to think about different systems for the production of inverted populations in the infrared and visible regions. We shall see in the following sections that many different approaches were considered almost simultaneously and independently, and that, in the case of semiconductors, consideration of the possibility of producing radiation through stimulated pair recombination even preceded the Schawlow and Townes discussion.

Of course the main lines were influenced by the ideas of these two researchers and most people were expecting the first laser action to take place in an excited gas. However, some researchers were thinking along completely different lines about the type of resonant cavity to be used, showing that in some sense the time was ripe anyway for the construction of laser devices (then still known as optical masers).

6.2 The ruby laser

The first laser was realised in July 1960 at the Hughes Research Laboratories, Malibù (Southern California) by T H Maiman, using ruby as the active material.

Theodore H Maiman (1927-) was the young head of the quantum electronics division at Hughes. He had obtained his BSc in engineering physics from Colorado University and his MSc in electrical engineering and PhD (1955) in physics from Stanford University, where his doctoral research was in microwave spectroscopy. Then he became a research scientist at Lockheed Aircraft for a short while, studying communication problems connected with guided missiles.

In his first five years at Hughes Maiman worked on masers. In fact, he developed the first ruby maser cooled with liquid nitrogen, and then one cooled by dry ice[1].

Speculations about laser materials during the first half of 1960 centred around gases, specifically optically excited alkali vapours and noble gases excited in an electric discharge. Maiman's achievement of the ruby laser came as a surprise, but it was not an accidental discovery. Having worked with ruby as a (microwave) maser material for some time, Maiman had studied the spectroscopy of the chromium ion in pink ruby, and his construction of the laser was preceded by an accurate spectroscopy study of fluorescence, received by *Physical Review Letters* on 22 April 1960 and published in June 1960[2]. He also gave a simplified energy-level diagram for triply ionised chromium in corundum (figure 6.1)[3]. When the ruby was irradiated with light at a wavelength of about 5500 Å (green) chromium ions were excited to the $^4F_2$ state and then quickly lost some of their excitation energy through non-radiative transitions to the ^2E state. This state then slowly decays by spontaneously emitting a sharp doublet, the components of which at 300 K are at 6943 Å and 6929 Å (the R doublet). The lifetime of this process was determined by Maiman to be about 5 ms. This rather long remanence time of atoms in the metastable state ^2E, and the successive decay to the fundamental state, are responsible for the fluorescence phenomenon in ruby. In addition, the decay rate S_{32} from the $^4F_2$ to the doublet ^2E was found to be $S_{32} \simeq 2 \times 10^7\ s^{-1}$.

A measurement of fluorescent quantum efficiency, i.e. the number of fluorescent quanta emitted compared to the number absorbed by the crystal from the exciting beam, yielded a value close to unity.

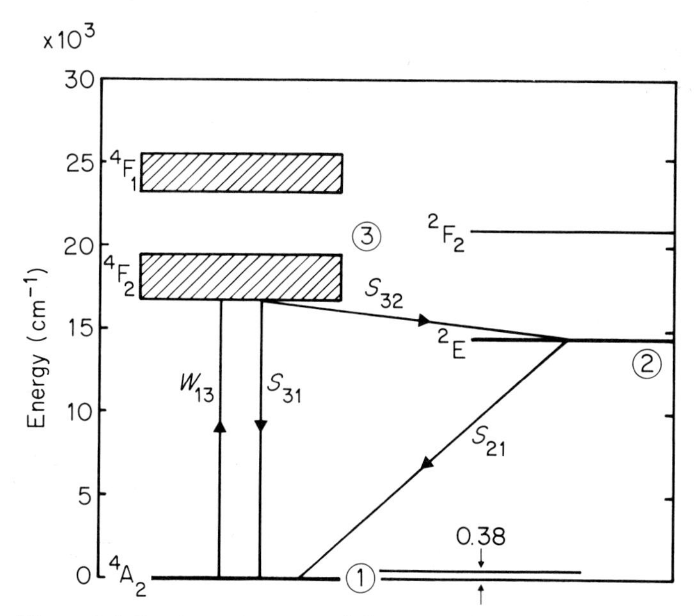

Figure 6.1 Pertinent features of the energy level diagram in ruby[3].

Maiman calculated that population changes could easily be produced in the $^4A_2$ ground state of ruby under suitable optical excitation. He measured them under the influence of 5600 Å radiation. This was done by observing changes in the magnitude of the 11.3 GHz absorption due to zero-field ground-state splitting under illumination with a short light pulse from a flash tube, and the 4100 Å optical absorption ($^4A_2 \rightarrow {}^4F_1$ transition) resulting from 5600 Å excitation.

When an intense pulse of radiation at 5600 Å was turned on, the 4100 Å radiation passing through the crystal abruptly increased and subsequently decayed in about 5 ms. This result was fairly well explained by the temporary reduction in the ground-state population, and both experiments allowed him to calculate that a population change of about 3% was obtained.

Maiman was working with pink ruby with a concentration of approximately 0.05% by weight of Cr_2O_3 to Al_2O_3. In a few months, by applying a sufficient and rapid irradiation from a xenon flashlamp, he obtained inversion of population and laser emission between the ^2E level and the ground state. Maiman's discovery was made public through a press announcement (*New York Times*) on 7 July 1960.

The paper containing his first results was, in fact, refused by *Physical Review Letters*[4] for the reason that maser physics was believed to have reached the stage where further advances no longer merited rapid publication. Moreover, people were confident in Schawlow's assertion that R lines in ruby were not suitable for laser action. However, Maiman's paper was soon published in British journals[5, 6] instead. In the 6 August issue of *Nature*[5] he described the experiment simply as follows:

> ... a ruby crystal of 1 cm dimensions, coated on two parallel faces with silver, was irradiated by a high-power flash lamp; the emission spectrum obtained under these conditions is shown in [figure 6.2(*b*)]. These results can be explained on the basis that negative temperatures were produced and regenerative amplification ensued.

In the paper in *British Communications and Electronics*[6] the shortening of the decay time was also discussed.

A detailed account of the experiment was published the following year in two papers in *Physical Review*[7, 8]. The first paper contained a full theoretical treatment in which, by using the rate equation formalism, the pumping power for the threshold condition was derived. The general class of materials considered was that of fluorescent solids whose emission spectra consist of one or more sharp lines. The excitation mechanism was considered to be by radiation of frequencies which produce absorption into one or more bands. Some of this excitation energy is lost by a combination of spontaneous emission and thermal relaxation to lower-lying states; however, if the solid has a relatively high fluorescent

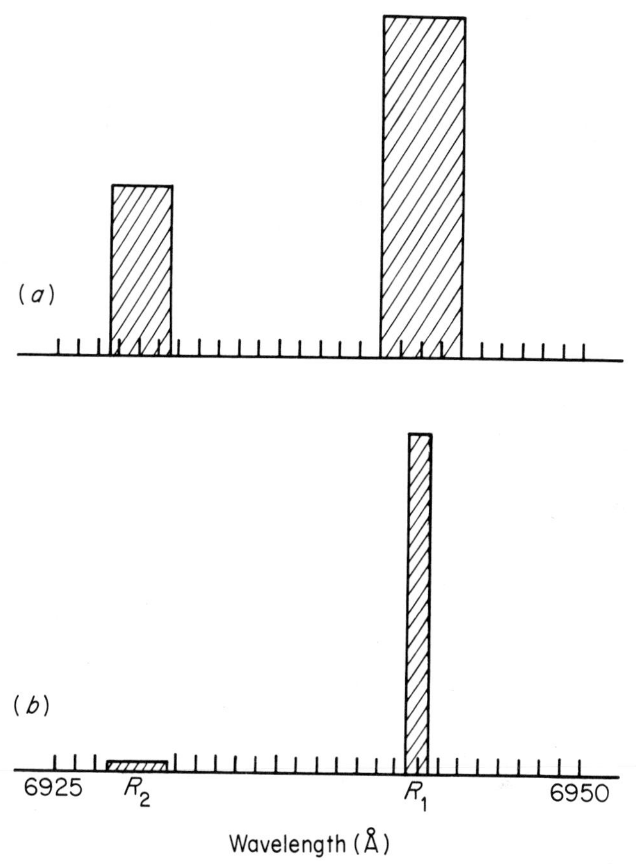

Figure 6.2 Emission spectrum of ruby: (*a*) low-power spectrum, (*b*) high-power excitation[5]. Reprinted by permission from *Nature* **187**, 493; copyright © 1960 Macmillan Journals Limited.

efficiency most of the energy is transferred to the sharp fluorescent levels by means of a non-radiative process. Subsequently, by a combination of spontaneous emission and thermal relaxation, the excited atoms (ions) return either to the ground state, or another low-lying state. The spontaneous emission from these sharp levels is the observed fluorescent radiation. If the exciting radiation is sufficiently intense it is possible to obtain a population density in one of the fluorescent levels greater than that of the lower-lying thermal state. In this situation, spontaneously emitted (fluorescent) photons travelling through the crystal stimulate upper-state atoms to radiate, and a net component of induced emission is superimposed on the spontaneous emission.

Three-level and four-level schemes were discussed (see figures 6.3(*a*) and (*b*), which are figures 1 and 2 of the original paper[7]). One of the features of the

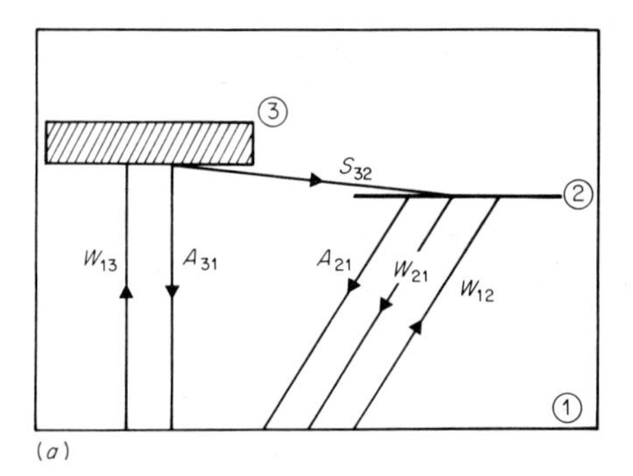

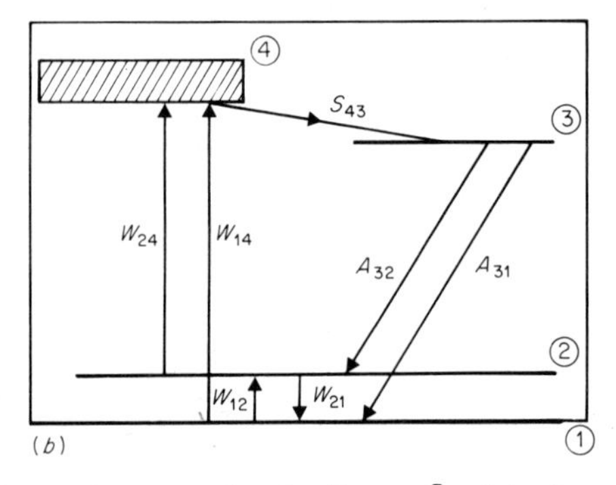

Figure 6.3 Optical energy-level diagram[7]: (*a*) for a three-level fluorescent solid, and (*b*) for a four-level system.

solid-state system was the possible use of a broad absorption band for the pump transition. This situation allows a relatively high pumping efficiency to be realised, since most high-power optical sources have very broad spectral distributions in their radiation energy. This overcame one of the particular problems foreseen by Schawlow and Townes.

In the three-level scheme the rate equations were written by Maiman as

$$dN_3/dt = W_{13}N_1 - (W_{31} + A_{31} + S_{32})\,N_3$$

$$dN_2/dt = W_{12}N_1 - (A_{21} + W_{21})\,N_2 + S_{32}N_3$$

$$N_1 + N_2 + N_3 = N_0,$$

where W_{13} is the induced transition probability per unit time for the transition $(1 \rightarrow 3)$ caused by the exciting radiation of frequency ν_{13}; W_{21} is the induced probability $(2 \rightarrow 1)$ due to the presence of radiation of frequency ν_{21}. A_{31} and A_{21} are Einstein spontaneous emission coefficients and S_{32} is the transition probability for the non-radiative process $(3 \rightarrow 2)$. N_1, N_2 and N_3 are respectively level population densities, and N_0 is the total active ion density in the crystal.

Thermal processes $(2 \rightarrow 1)$ and $(3 \rightarrow 1)$ were assumed to be negligible compared with the corresponding radiative processes, but that the reverse was true for the transition $(3 \rightarrow 2)$.

In the steady state, where the time derivatives are zero, an approximate solution was found to be

$$(N_2 - N_1)/N_0 \simeq (W_{13} - A_{21})/(W_{13} + A_{21} + 2W_{12}),$$

provided $A_{31} \ll S_{32}$, $N_3 \ll N_1$, $N_3 \ll N_2$. Therefore, by making $W_{13} > A_{21}$, stimulated emission can, in principle, be obtained. Considerations concerning losses led to the more restrictive condition that $N_2 - N_1$ must be sufficient to overcome circuit losses, that is

$$(N_2 - N_1)/N_0 \simeq (1 - r)/\alpha_0 l,$$

where r is the reflection coefficient of the silvered end plates, α_0 is the normally measured absorption coefficient for the transition $1 \rightarrow 2$ under low-power excitation, and l is the length of the material.

A similar analysis was carried out in the case of four levels. The use of rate equations of the kind written here, which we have already seen in the Bloembergen treatment of the solid-state maser, immediately became of wide and common use in all treatments of lasers.

The next argument to be dealt with in the paper concerned the spectral width. The presence of multimode oscillation was considered and the importance of strains, inhomogeneities and deviations from single crystallinity – which tend to scatter energy into undesired modes – was recognised. No discussion of coherence properties of the radiation was entered into, nor was there any consideration of the statistics of the radiation which was clearly produced in a non-equilibrium thermodynamic situation. Apparently, nobody was aware of this, and no discussion of this point can be found in any of the papers on lasers published during that time. Attention was focused essentially on the use of rate equations to calculate threshold pumping levels and, later on, spiking behaviour, Q-switching properties, etc.

The second paper was a description of the results of spectroscopic and stimulated emission experiments. In particular, spiking behaviour and line narrowing were observed. The various ruby samples were divided into two categories:

(1) Crystals which exhibited R_1 line narrowing by only about four or five times, had a faster but smooth output time decay (compared with the fluorescence), an output beam angle of about 1 rad, and no clear-cut evidence of a threshold excitation. Clearly these were highly strained and bad crystals;

(2) Crystals which exhibited pronounced line narrowing of nearly four orders of magnitude, an oscillatory behaviour of the output pulse, and a beam

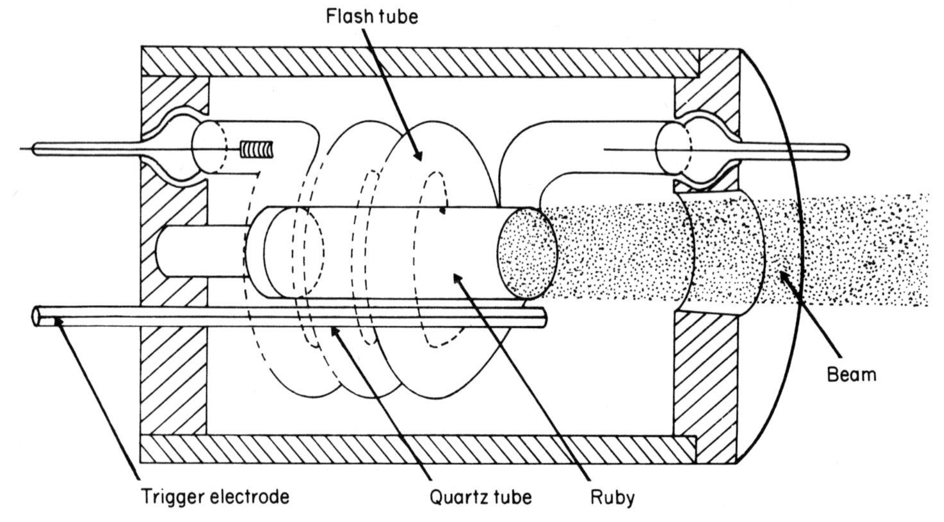

Figure 6.4 Apparatus used by Maiman for the first ruby laser[8].

angle of about 10^{-2} rad; these crystals were in particular characterised by a very clear-cut threshold input energy where this pronounced line and beam narrowing occurred.

The laser realised by Maiman is shown in figure 6.4 (figure 7 of the original paper[8]). Maiman wrote[8]

> Due to the need for high source intensities to produce stimulated emission in ruby and because of associated heat dissipation problems, these experiments were performed using a pulsed light source ... The material samples were ruby cylinders about $\frac{3}{8}$ inch in diameter and $\frac{3}{4}$ inch long with the ends flat and parallel within $\lambda/3$ at 6943 Å. The rubies were supported inside the helix of the flash tube, which in turn was enclosed in a polished aluminium cylinder [see our figure 6.4]; provision was made for forced air cooling. The ruby cylinders were coated with evaporated silver at each end; one end was opaque and the other was either semitransparent or opaque with a small hole in the center.

The population inversion was obtained between the level ^2E and the ground level and light emission was amplified by multiple reflection at the end mirrors.

In this first laser the line R_1 was active (figure 6.5). The energy for the flash tube was obtained by discharging a 1350 μF capacitor bank. The input energy was varied by changing the charging potential. Threshold was obtained at energies between 0.7 and 1.0 J according to the way the terminal faces of the rod were prepared.

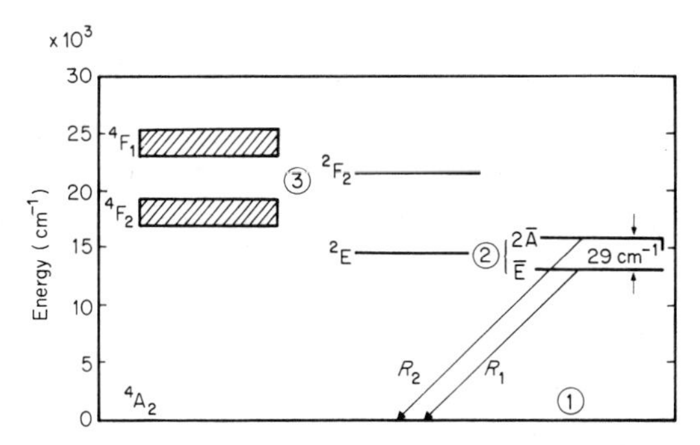

Figure 6.5 Energy-level diagram for ruby[8].

The ruby laser system was patented by Maiman (patent no. 3.353.115 issued 11/14/1967). Immediately after his realisation of the ruby laser at Hughes, Maiman left and first became president of Korad Co, a subsidiary of Union Carbide. He is now vice-president, Advanced Technology, TRW Electronics.

The principal characteristics of the laser beam, its coherence, directionality and collimation were studied immediately by Schawlow, together with some collaborators, at Bell Telephone Laboratories. In the *Physical Review* paper by Maiman[8] references are made to this and other papers showing the frantic activity in the field which immediately started after his announcement in June.

Schawlow *et al*, in a paper sent to *Physical Review Letters* in August 1960[9], verified the narrowing of the emission line, finding a linewidth of 0.2 cm^{-1} in lasing conditions as compared to 6 cm^{-1} in normal emission.

They also verified the sudden variation of the directional distribution of the emitted light when the threshold condition for excitation was overcome, and the presence of relaxation oscillations (figure 6.6). At that time a similar behaviour was known to occur at microwave frequencies in ruby masers[10] and was attributed by Statz and de Mars[11] to time-dependent interaction between the inverted population of the electron-spin systems of the paramagnetic substance and the resonant cavity.

In the case of ruby one possible explanation given in the Schawlow *et al* paper[9] was that the stimulated emission, once it sets in, proceeds at a rate greater than that at which atoms are being excited to the $\bar{E}$ (^2E) state. When this occurs, the stimulated emission may drive the inverted population below the level at which the process first sets in, so that, when the stimulated emisson is finally quenched, some definite time interval will be required for the negative population to be restored.

In this paper there is also the first investigation of spatial coherence of the laser (first-order coherence). Schawlow *et al* opened a rectangular aperture

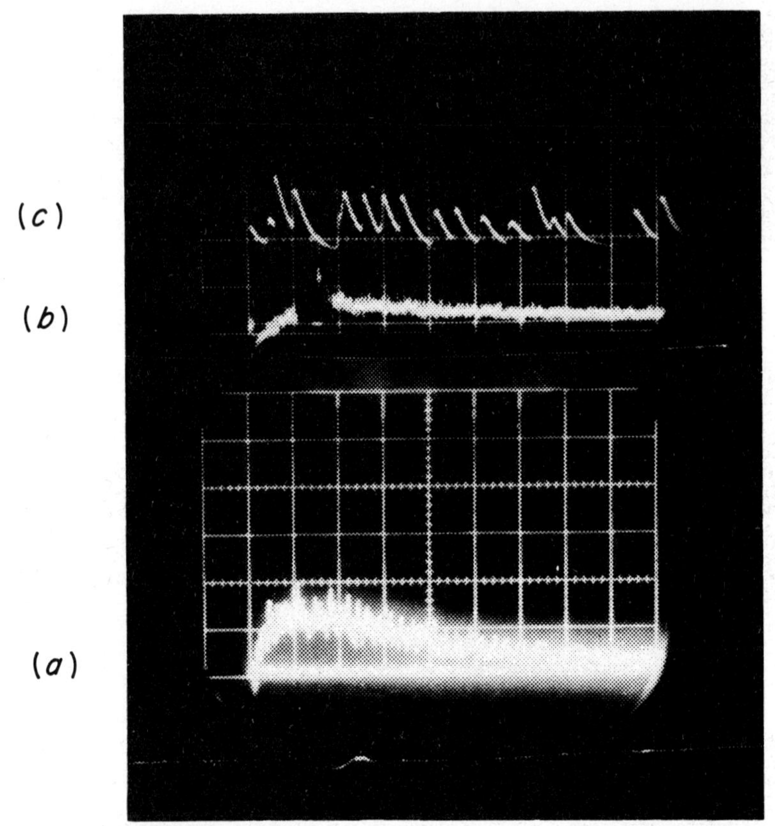

Figure 6.6 Spiking behaviour of a free-operating ruby laser; from Collins *et al*, *Phys. Rev. Lett.* **5**, 303 (1960); © The American Physical Society. The lower trace shows decay of fluorescence for a low excitation level. The trace in the middle shows lasing action at a higher excitation level when the trace, after 500 μs, breaks. The upper trace is an enlargement of 100 μs of the lasing trace, showing spikes.

50 × 150 μm in size in a heavily silvered coating of one end of the rod and looked for diffraction. The image was found to consist of a Fraunhofer diffraction pattern for a rectangular aperture illuminated by wavefronts which were approximately plane and approximately coherent. This pattern disappeared when the excitation was reduced below the threshold. Another interesting result of this experiment was that laser emission was taking place from small active regions on the face of the rods, each ≃500 μm in diameter. Later on this filamentary structure of the emission was intensively studied. The investigators also observed that a decrease in the excitation threshold of 30% could be obtained by lowering the working temperature.

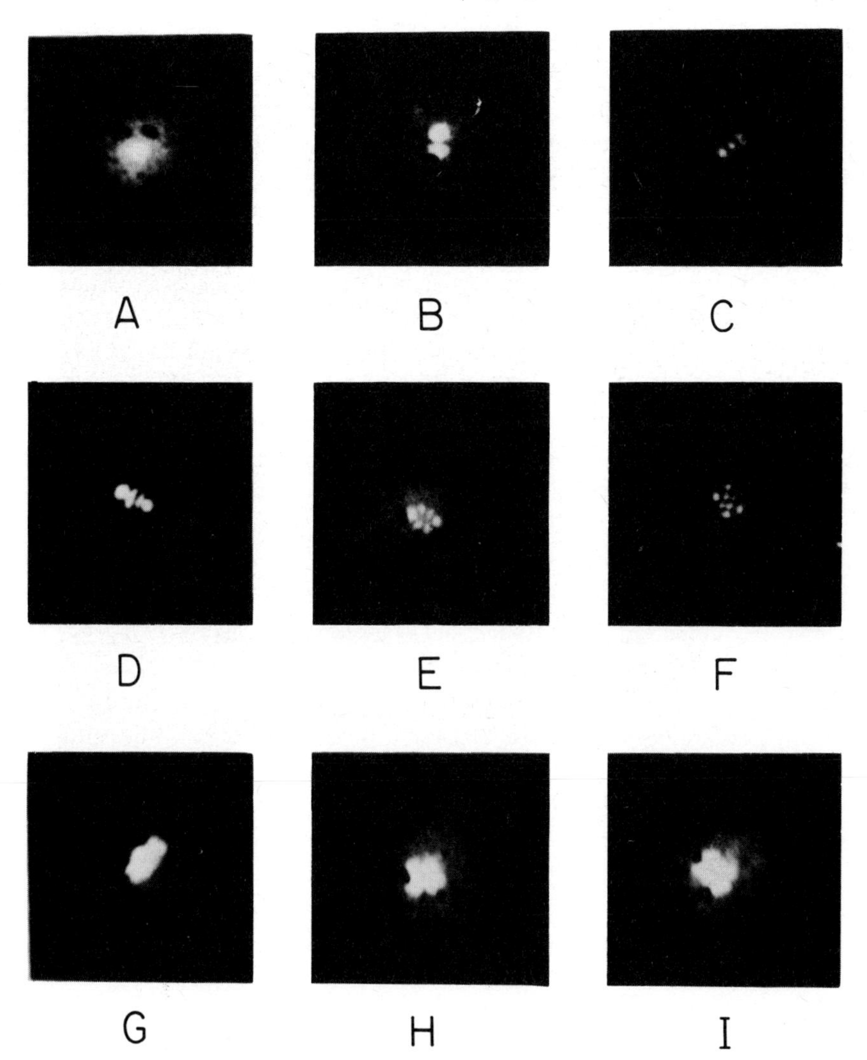

Figure 6.7 Typical patterns on the end of a ruby laser. (From V Evtuhov and J K Neeland in *Quantum Electronics* vol 3, edited by P Grivet and N Bloembergen (Columbia University Press: New York, 1964) p1405.)

Further studies were made of the first-order spatial and temporal coherence of the emission[12], and mode structure was observed by Evtuhov and Neeland[13] (figure 6.7). Longitudinal modes were studied through their beats by Siegman[14]. Direct studies of longitudinal modes were made using a high-resolution, high-dispersion grating spectrometer both by Duncan *et al*[15] and Ciftan *et al*[16].

Schawlow[17] had originally proposed that the so-called N lines in ruby be used[18], since these transitions have a somewhat lower state above the ground state, and can therefore be depopulated at low temperature. Shortly after the announcement of laser action at the R_1 transition, he and Devlin[19] and also Wieder and Sarles[20] reported the observation of laser action at 7009 and 7041 Å, which are two of the strongest N lines. Emission on the R_2 line was obtained in 1962 by McClung, Schwarz and Meyers[21].

The transverse mode emission was strongly influenced by the optical quality of the ruby[22]. The original ruby laser material was grown by the Verneuil flame fusion technique[23], a process long used in both Europe and the United States for gems and bearings, which had been developed in 1904. The Czochralski process was applied later[24]. Improvement of ruby laser characteristics was obtained by using better crystals, and more efficient light pumping schemes[25].

A continuously operating ruby laser was developed by Nelson and Boyle[26] in 1962 by using an arc lamp and a trumpet-shaped laser rod with a special arrangement of mirrors to image the lamp upon the end of the ruby.

In the early days of research it had already been shown that by focusing ruby laser light it was possible to vaporise any kind of materials. Very soon people began to drill holes in razor blades by focusing ruby lasers on to them and someone suggested that the laser output could be measured in Gillettes!

Schawlow became a well trained demonstrator of the special properties of laser light, aiming the ruby beam on to a blue balloon contained in a larger red one. The red ruby light was not absorbed by the red balloon, which was unaffected, but made the blue balloon explode because it absorbed the light. The experiment was perfected by building a small, portable ruby laser into the housing from a toy ray-gun.

The first demonstration was given at the meeting of the American Association for the Advancement of Science at Cleveland in December 1963, and Schawlow still uses it to show how this exercise eventually led him to obtain a patent for a laser eraser, which demonstrates the more amusing paths to discovery!

6.3 The four-level laser

A few months after the realisation of the ruby laser, a four-level laser at low threshold was obtained by Sorokin and Stevenson[27] by using trivalent uranium in calcium fluoride.

In the four-level laser the population inversion is more easily obtained (figure 6.3(*b*)). In this case the critical population inversion between levels 3 and 2 can be obtained simply by maintaining a population in level 2 equal to the theoretical minimum.

Sorokin and Stevenson observed emission at 2.5 μm in CaF_2 crystals with 0.05% molar concentration of uranium. Later, with 0.1% concentration, they observed emission at 2.6 μm[28].

Considerable confusion existed regarding the origin of these lines and the identification of the levels involved in the fluorescence of U in calcium fluoride.

Although an extensive literature on optical properties of ions in solids existed at that time[29], most of the spectral information needed to develop these lasers still had to be obtained while the research was in progress. A group of investigators at Bell Telephone Laboratories undertook a detailed study of the energy levels of U in CaF_2. By means of paramagnetic resonance techniques they showed, after some uncertainty[30], that the U ion may be located in CaF_2 in tetragonal[31], trigonal, and orthorhombic sites. Emission at 2.51 μm and at 2.44 μm was assigned to U^{3+} in tetragonal sites. Orthorhombic U^{3+} is responsible for emission at 2.61, 2.24 and 2.57 μm, and emission at 2.24 μm was observed in crystals containing 90% of U^{4+} ions in trigonal sites[32].

The experiments of Sorokin and Stevenson were performed at liquid helium temperature; later Bostick and O'Connor[33] reported operation at 77 K, and Miles[34] at up to 300 K. In the early experiments flash excitation was used, and regular relaxation oscillations, as in ruby, were observed. Continuous operation was obtained by Boyd *et al*[30] at 77 K, with the arrangement shown in figure 6.8.

Meanwhile, in 1961, Sorokin and Stevenson[35] obtained stimulated emission in divalent Sm in CaF_2 at a wavelength of 7083 Å. In the same year Johnson and Nassau[36] reported stimulated emission of Nd^{3+} in $CaWO_4$ at 1.06 μm, and a few months later they reported continuous operation at ambient temperatures using the same equipment as shown in figure 6.8[37]. In the first experiment the operation was obtained for a few seconds, but when the problem of charge compensation was understood and the number of available levels was reduced by the addition of Na to the crystal, continuous operation became possible for periods up to an hour[38].

The next lasing materials were Tm and Ho[39], still in $CaWO_4$. In 1962 the Bell Laboratories group obtained emission from Nd^{3+} in a number of different host crystals[40].

As is evident, attention was focused on rare-earth ions (trivalent and divalent rare-earth ions), actinide ions and transition-metal ions. Somewhat bitterly, Lengyel writes[41]:

> Systematic and exhaustive description of four-level lasers is a difficult task because of the large number of combinations possible with a variety of active ions embedded in a variety of host lattices. The compilation of information in this field is further hindered by investigators whose aim is to maximise the number of their publications. The four-level lasers field is notorious for the fact that separate papers were written about stimulated emission in several ion-host-lattice combinations even when the same apparatus was used for making the measurements and when the papers had to appear in the same issue of a journal.

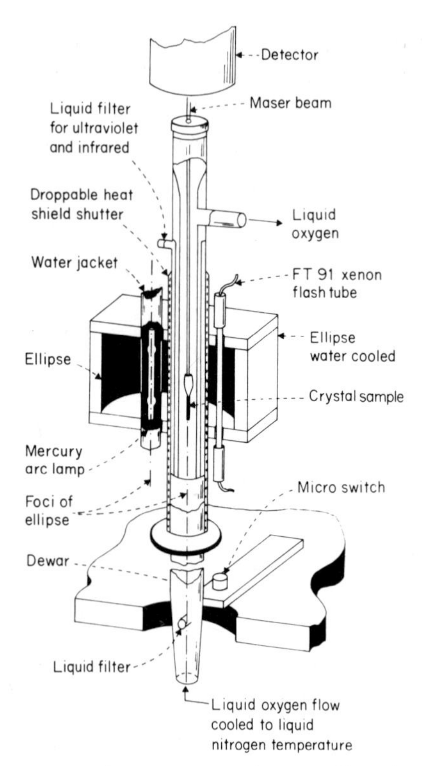

Figure 6.8 Experimental set-up used at Bell Telephone Laboratories[37]. Copyright © 1962 IEEE.

In a paper on rare-earth ion lasers at the beginning of 1963 L F Johnson[42] quoted laser action in six trivalent rare-earth ions in various host materials (Nd^{3+}, Pr^{3+}, Tm^{3+}, Ho^{3+}, Er^{3+}, Yb^{3+}) and three divalent ions (Tm^{2+}, Dy^{2+}, Sm^{2+}). Laser action was observed in ten materials containing Nd^{3+}. Energy levels of trivalent rare-earth free ions were tabulated by Dieke and Crosswhite[43].

Meanwhile, considerable interest had arisen over the use of garnets as magnetic device materials and the possibility of rare-earth substitution into the yttrium garnets was demonstrated. A survey of Nd-doped yttrium aluminium, yttrium gallium and yttrium gadolinium garnets, by Geusic *et al*[44] showed that the most promising material was YAG; and today the only really satisfactory solid-state continuous-wave laser has Nd^{3+} in YAG as the active medium[45].

In 1961 Snitzer[46] obtained laser action by neodymium doping of crown glass in the form of cylindrical rods or fibres. Over 100 J of energy were obtained from such rods in a single pulse[47].

There are a number of characteristics which distinguish glass from other solid laser host materials. Its properties are isotropic. It can be doped at very high concentration with excellent uniformity. It is a material which affords considerable flexibility in size and shape, and may be obtained in large homogeneous pieces of diffraction-limited optical quality. It can also be relatively cheap in large volume production and can be fabricated by a number of processes, such as drilling, drawing, fusion and cladding, which are generally alien to crystalline materials. Continuous laser action at 1.06 μm was obtained at room temperature in a barium crown glass doped with Nd[48]. Laser action in glasses was obtained subsequently with Yb^{3+}, Er, Ho, and Gd[49].

6.4 The gas laser

Although gas lasers were not operational until after the realisation of the ruby laser, gaseous systems had been considered since the beginning. The use of a gas discharge for obtaining population inversion was proposed by Fabrikant[50] as early as 1939. He suggested the use of a buffer gas to shorten the relaxation time of the lower laser level through collisions of the second kind. Also A Ferkhman and S Frish claimed to have realised population inversions[51] way back in 1936.

After Schawlow and Townes[52] had considered selective optical pumping in gases, proposals for obtaining, through selective excitation, the inversion of the upper laser level in a one-gas system were made in 1959 by Sanders[53] and Javan[54], who both suggested producing the required population inversion by using an electric discharge in a gas.

Sanders did his work while he was at Bell Laboratories, on leave from the Clarendon Laboratory, Oxford, England. In a very short letter to *Physical Review Letters* he observed that it was difficult to obtain a sufficient number of excited atoms with a flash lamp, and suggested the use of excitation produced by electron collisions. Such an excitation could easily be produced in an electrical discharge in a gas or a vapour. Population inversion could be produced if excited states with a long lifetime existed in the active material and there were states present at lower energies with a short lifetime. He observed, too, that in an electrical discharge many processes exist which:

> ... may disturb the relative population of such levels from the value implied by these lifetime values, and it may be possible to choose conditions which maintain a high population in the upper state.

In 1950 W E Lamb[55] had already noted that electron collisions in a gas discharge can create an inverted population, as we have seen in Chapter 2.

The very next paper in the same issue of *Physical Review Letters* was written by A Javan[54], who too had considered these problems. Javan realised that any single physical process tends to produce a Boltzmann distribution; therefore a medium with population inversion can be produced in a steady-state process only as a result of the competition of several physical processes proceeding at different rates. After having discussed pure gases, he then focused attention on certain kinds of gas mixtures.

The following exchange processes were considered:

(1) electron collision of the first kind in which an atom gains energy from an electron;

(2) electron collision of the second kind in which an excited atom loses energy to an electron;

(3) spontaneous emission of radiation from an excited atom;

(4) absorption of radiation by an atom;

(5) stimulated emission of radiation by an atom.

Javan used θ_{ij} to indicate the lifetime of the transition of an atom from level i to level j, when the atom is subjected only to collisions with electrons of a given density in equilibrium among themselves at the absolute temperature T. If no other processes than these collisions were to take place, the rate of change of the number of atoms at level i would be given by

$$\mathrm{d}N_i/\mathrm{d}t = \sum_j [(N_j/\theta_{ji}) - (N_i/\theta_{ij})]. \tag{6.1}$$

If we assume that thermodynamic equilibrium is established at temperature T, the number of atoms in each level will be stationary. Then

$$\sum_j [(N_j^*/\theta_{ji}) - (N_i^*/\theta_{ij})] = 0, \tag{6.2}$$

where N^* is the stationary value of N.

The principle of detailed balance now requires that the exchange between each pair of energy levels should balance out. This means that not only must the sum in equation (6.2) vanish, but that

$$(N_j^*/\theta_{ji}) - (N_i^*/\theta_{ij}) = 0 \tag{6.3}$$

must hold for every i and j. Therefore

$$\theta_{ij}/\theta_{ji} = N_i^*/N_j^* = (g_i/g_j) \exp[-(E_i - E_j)/kT] \tag{6.4}$$

and equation (6.4) is always valid.

Now let us consider the simplest case of two levels, with a spontaneous radiative transition, with a lifetime τ_2 from levels 2 to 1; the rate equation is

$$\mathrm{d}N_2/\mathrm{d}t = (N_1/\theta_{12}) - (N_2/\theta_{21}) - (N_2/\tau_2). \tag{6.5}$$

In the stationary state

$$N_2/N_1 = (1/\theta_{12})/[(1/\theta_{21}) + (1/\tau_2)]. \tag{6.6}$$

When the radiative process is comparatively fast, so that $\tau_2 \ll \theta_{21}$, we have

$$N_2/N_1 \simeq \tau_2/\theta_{12} = (g_2\tau_2/g_1\theta_{21}) \exp[-(E_2-E_1)/kT]. \tag{6.7}$$

The ratio τ_2/θ_{21} is a measure of the departure from the Boltzmann equilibrium distribution.

To some extent this factor is under the control of the experimenter, since τ_2 is fixed; but $1/\theta_{21}$ is proportional to the electron density in the discharge. However, N_2/N_1 cannot be increased arbitrarily by increasing the electron density, because the validity of equation (6.7) is predicted on the assumption $\tau_2/\theta_{21} \ll 1$.

However, if a third level, 3, is introduced above level 2, and if we assume no direct interaction between levels 2 and 3, we find, under similar assumptions,

$$N_3/N_1 \simeq (g_3/g_1)(\tau_3/\theta_{31}) \exp[-(E_3-E_1)/kT]. \tag{6.8}$$

In the case of equal multiplicities, we conclude from equations (6.7) and (6.8) that

$$N_3/N_2 \simeq (\tau_3/\tau_2)(\theta_{21}/\theta_{31}) \exp[-(E_3-E_2)/kT]. \tag{6.9}$$

This equation was given in the paper by A Javan, without the detailed discussion we have presented here. Javan argued that, even though the exponential factor is always less than one, the factor in front of the exponential may be sufficiently larger than one for certain values of the parameters to cause $N_3/N_2 > 1$.

Javan considered that the $3\,^1D$ and $2\,^1P$ levels of He have a lifetime ratio $\tau_3/\tau_2 = 35$ and an estimate of θ_{31}/θ_{21} gave 15.

Equation (6.9) indicates then that a negative temperature may be possible. Unfortunately, the collision cross section for excitation of $3\,^1D$ is relatively small. At a pressure of about 5×10^{-3} mm of Hg the relatively high fraction of 1% of the total number of atoms would have to be ionised to give the 10^9 excited atoms per unit volume needed to overcome resonator losses and exceed the threshold conditions for oscillation.

Javan also considered Ne, and, to prevent an unfavourable population of the excited levels, suggested the introduction of a very small amount of a quenching gas, such as argon, to decrease the lifetimes of the longer-lived excited atoms.

An alternative scheme was finally suggested, which used the transfer of excitation between excited states of two different atoms in a gaseous mixture. Javan wrote:

> Consider a long-lived state of an atom (such as a metastable state). This state can be populated appreciably at moderate electron densities.
>
> If an excited state of a second atom happens to lie very close in energy to that of the level of the first atom, a large cross section is expected to exist for an inelastic collision resulting in a transfer of excitation from the

metastable state to the excited state of the other atom and vice versa.

Due to the non-adiabatic nature of the process of collision, the levels of the second atom which differ in energy considerably from that of the metastable level of the first atom do not show appreciable cross sections for transfer of excitations.

A few other considerations then followed and mixtures of Kr-Hg and He-Ne were proposed.

The paper was received at *Physical Review Letters* on 3 June 1959. The ideas contained therein were also presented at the *Quantum Electronics Conference*[56]. Javan knew Ladenburg's work quite well, as is shown by a quotation he made in the *Phys. Rev. Lett.* paper of the *Rev. Mod. Phys.* paper by Ladenburg (Chapter 2 note 44).

A few months later, Javan with a few other colleagues at Bell succeeded in realising a gaseous laser in a He-Ne mixture. On 14 December 1960, by using a Kerr cell, a telephone conversation took place at Bell Laboratories. The He-Ne gas laser was presented to the press for the first time on 31 January 1961 by A Javan and W R Bennett jr of the Physical Research Department and D R Herriott of Systems Research at Bell Laboratories. The paper describing the laser was received by *Physical Review Letters* on 30 December 1960 and published in the February 1961 issue[57].

The laser was realised with a radiofrequency discharge in a mixture of He-Ne. Five different wavelengths in the infrared were active, giving continuous operation. Population inversion was achieved between several Ne levels by means of excitation transfer from the metastable He (2^3S) to the 2s levels of Ne (see figure 6.9)[58]. Although the $2s_2$ and $2s_4$ levels of Ne may radiate to the Ne ground state, in the limit of complete resonance trapping their lifetimes are determined primarily by radiative decay to the 2p levels. Under these conditions, the lifetimes of all of the 2s levels are about one order of magnitude longer than those of the corresponding 2p states. (The decay of the 2p levels is due to their radiative decay to the 1s levels). Thus population inversions may be obtained on each of the 30 allowed $2s \rightarrow 2p$ transitions.

The continuous-wave oscillation was observed in a discharge containing 0.1 mmHg of Ne and 1 mmHg of He in a quartz tube with an inside diameter of 1.5 cm and length 80 cm terminated with 13-layer evaporated-dielectric-film plane mirrors placed inside the tube.

Oscillations at 11180, 11530, 11600, 11990 and 12070 Å were observed. The linewidth of the 11530 Å line (15 mW – the strongest one) was measured using the beating technique in a photomultiplier. The measured linewidth was in the range of 10–80 kHz and the angular spread of the beam was less than one minute of arc. Evidence of modes was obtained. The emitted beam had a diameter of 0.45 inch (~ 1 cm).

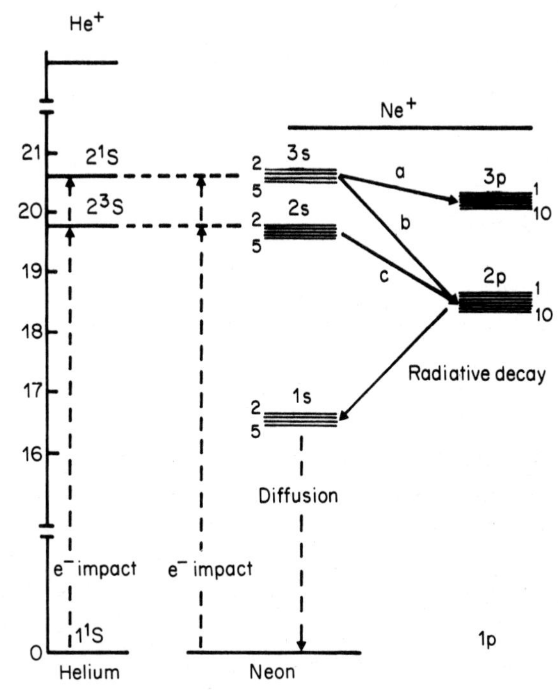

Figure 6.9 Energy levels of helium and neon. The helium levels are designated by the Russell–Saunders notation and the neon levels by Paschen notation. The laser transitions in the infrared (a = 2.8–4.0 μm, c = 1.1–1.5 μm) and visible (b = 0.59–0.73 μm) are shown.

A complete account of the experimental realisation was given by Donald R Herriott at the Optical Society of America Spring Meeting at Pittsburg, March 1961 and at the Second International Conference on Quantum Electronics, Berkeley, 23–25 March 1961, and published a few months later in the *Journal of the Optical Society of America*[59]. Figure 6.10 (figures 1 and 5 of this paper) shows the first realisation and its mounting. The paper also studied several properties of the laser and the importance of this laser for the understanding of many lasing properties (hole burning, mode patterns[60], etc) became immediately apparent.

Visible output at 6328 Å was obtained in 1962[61], by changing the reflectivity of the external mirrors and later operation at 3.39 μm[62] in the IR was achieved. The pertinent levels are also indicated in figure 6.9. The 3.39 μm line shares a common upper state with the 6328 Å line. This means that they compete. Moreover, the 3.39 μm line has a much greater gain than that on the 6328 Å line, and in tubes a metre or more in length the infrared emission is strongly superradiant. This is the reason why the gain on the 6328 Å line does not increase beyond a certain limit when the length of the discharge tube is increased.

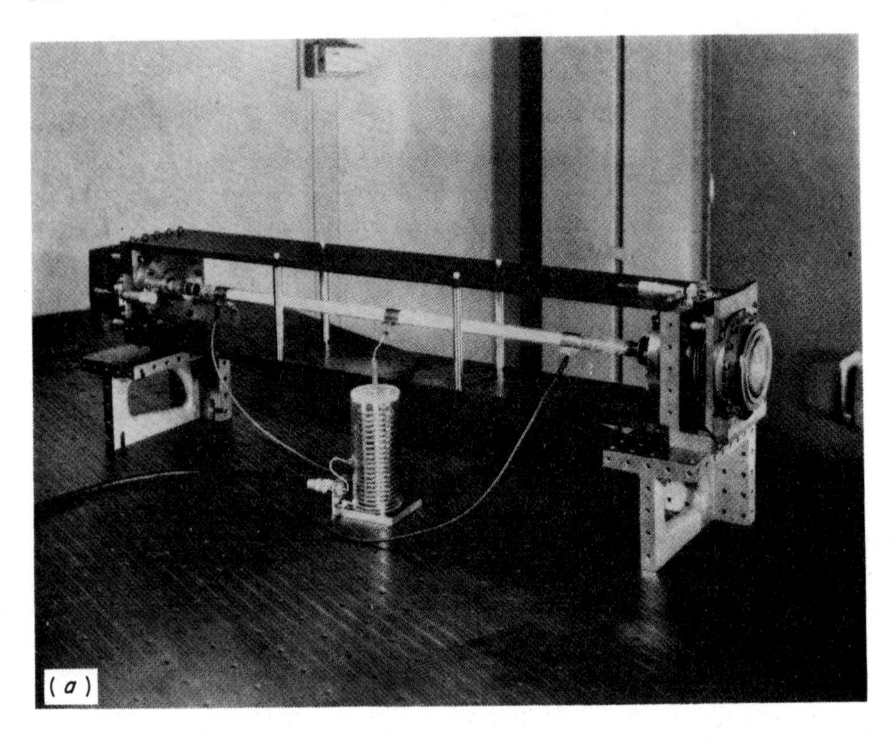

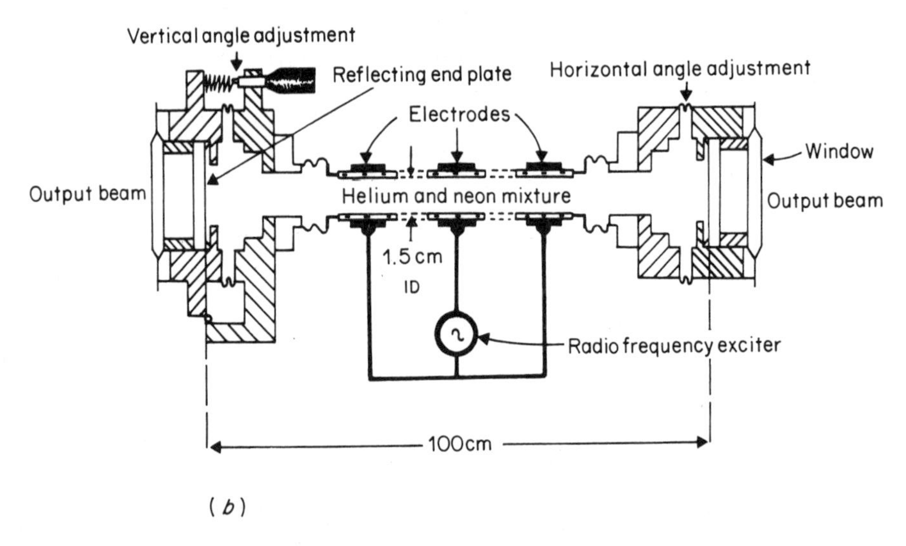

Figure 6.10 (*a*) The first mounting for the He–Ne laser, (*b*) diagram of the laser. (From D R Herriott, *J. Opt. Soc. Am.* **52**, 31 (1962); reprinted with permission from the Optical Society of America.)

The kinetics of the processes of interaction between the atoms of the two gases was developed by Basov and Krokhin[63, 64]. The two gases were distinguished by the letters a and b. The atoms of the working gas a, had three relevant levels; the atoms of the auxiliary gas b, had two relevant levels, and the upper level of the two gases was coincident or nearly so (figure 6.11).

The rate of change of the number of atoms of gas a on the upper energy level E_3 was written as

$$dN_3^a/dt = N_1^a\,[(1/\theta_{13}) + (1/t_{ba})] - N_3^a\,[(1/\theta_3) + (1/t_{ab}) + (1/\tau_3)] \quad (6.10)$$

where $1/t_{ba}$ is the rate per ground-state atom of gas a of resonant transfer of energy in collisions with atoms of gas b on level 3; $1/t_{ab}$ is the rate per excited atom of gas a, for the inverse process; $1/\theta_{13}$ is the rate of excitation per atom from level E_1 to level E_3 by electron collisions of the first kind and $1/\theta_3$ is the total transition rate per atom of gas a from level E_3 resulting from collisions of the second kind with electrons; and $1/\tau_3$ is the total transition rate from level E_3 arising from the remaining relaxation process. Excitation from level 2 to level 3 was neglected in this analysis because a term describing the process is proportional to N_2^a, and hence much smaller than the leading terms in equation (6.10).

The rate of change of the number of atoms N_2^a on level E_2 was written

$$dN_2^a/dt = (N_1^a/\theta_{12}) + N_3^a\,[(1/\tau_{32}) + (1/\theta_{32})] - N_2^a\,[(1/\theta_{21}) + (1/\tau_2)], \quad (6.11)$$

where $1/\theta_{12}$ is the rate of excitation by electrons; $1/\theta_{21}$ is the rate for the inverse process; $1/\theta_{32}$ is the transition rate from level E_3 to level E_2 arising from electron collisions; and the τ indicate radiative lifetimes.

In the stationary case $dN/dt = 0$ and the condition for the existence of a population inversion between levels E_3 and E_2 is

$$N_3^a/N_2^a > 1.$$

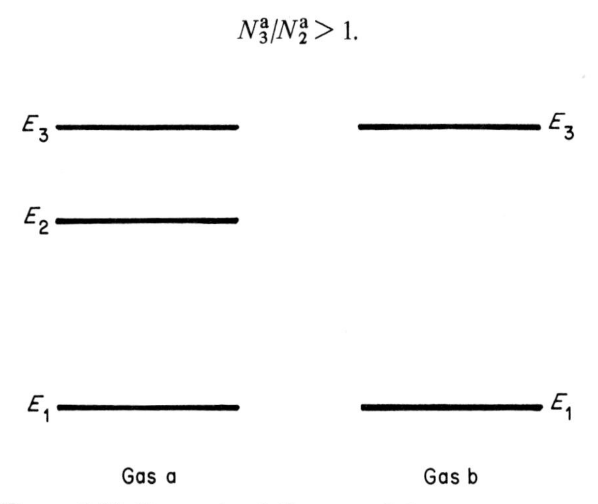

Figure 6.11 Energy-level diagram of the two-gas system.

According to collision theory, $t_{ba}/t_{ab} = N_1^b/N_3^b$ and assuming a Maxwellian distribution of electrons, or

$$\theta_{31}/\theta_{13} = \exp[-(E_3 - E_1)/kT],$$

the previous inequality can be written as

$$(\theta_{13}/t_{ba})\{1 - (N_1^b/N_3^b)\exp(-E_2/kT) + \theta_{21}[(1/\tau_{21}) - (1/\tau_{32}) - (1/\theta_{32})]\}$$
$$> \{\theta_{31}[(1/\theta_3) + (1/\tau_3)]\exp[(E_3 - E_2)/kT] - 1$$
$$+ \theta_{21}[(1/\tau_{32}) + (1/\theta_{32}) - (1/\tau_{21})]\}, \qquad (6.12)$$

where the ratio N_1^b/N_3^b is to be considered as an external parameter, i.e. the effective excitation temperature for atoms of the gas b will be considered as given.

The quantities θ_{31}/θ_3 and θ_{21}/θ_{32} do not depend on the density of the electrons in the discharge, but are determined by the cross sections of the corresponding processes and by the temperature of the electrons.

The inequality (6.12) has a different meaning in different regions of the variables $\theta_{21}[(1/\tau_{21}) - (1/\tau_{32})]$ and θ_{31}/τ_3. The range of these variables is divided by two curves into four regions. The equations of the curves are obtained by setting the expression in square brackets and the right-hand side equal to zero. In a first approximation $1/\theta_{32}$ may be neglected in comparison with $1/\tau_{32}$ and one may take $\theta_{31}/\theta_3 \simeq 1$; the curves then become straight lines of equations (figure 6.12),

$$\theta_{21}[(1/\tau_{21}) - (1/\tau_{32})] = N_1^b/N_3^b \exp(-E_2/kT) - 1, \qquad (6.13)$$

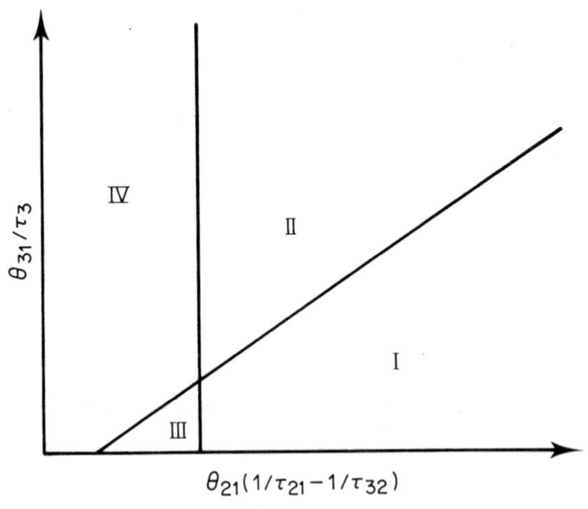

Figure 6.12 The Basov and Krokhin diagram.

$$\theta_{31}/\tau_3 = \{\theta_{21}[(1/\tau_{21}) - (1/\tau_{32})] + 1 - \exp[(E_3 - E_2)/kT]\}/\exp[(E_3 - E_2)/kT]. \quad (6.14)$$

The four regions in the figure therefore correspond to:

Region I, where $t_{ba} = \infty$ leads to population inversion in gas a, and gas b is unnecessary;

Region II, where population inversion occurs because of the presence of gas b;

Region III, where gas b impedes the formation of a population inversion; and

Region IV, where the occurrence of population inversion is impossible.

Although this analysis did not in fact even predict any possible laser system, it proved valuable in giving a fuller understanding of why certain gas lasers work.

Experimental research into the gas laser was, at the time these calculations came out, most intense. In his paper Javan[57] presented his first experimental realisation of the He–Ne laser, and mentioned concomitant research by Ablekov *et al*[65], writing: 'Recently some evidence for the presence of inverted populations in a Hg–Zn mixture has been reported ...'

In their paper Ablekov *et al* reported the observation of negative temperature or optical gain on the $4\,^1D_2 \rightarrow 4\,^1P_1^o$ transition of Zn at 6362 Å in a DC discharge containing Hg and Zn. Ablekov also claimed to have obtained line narrowing[66]. These results have never been repeated and no laser action has subsequently been obtained. However, these experiments show how the idea of utilising a discharge in a two-component gas was common to several research groups independently.

During the next year or so, more than 40 different transitions were obtained from the visible (0.63 μm) to the medium infrared (12 μm) by using ten different gaseous systems and at least four different excitation mechanisms[67].

The continuous operation obtained made it simpler to align the optical cavities, and many different kinds of resonators were considered. Meanwhile, theoretical treatments of modes in resonators were becoming available, and it may be useful at this point to summarise the situation in this respect.

6.5 Laser cavities

It is of importance to look carefully at the role of resonant cavities in laser design and evolution, because of the strong connection between the development of resonators and of the different kinds of lasers which took place at this time, after the first ruby and He–Ne lasers.

We have already noted that the use of a Fabry-Perot type cavity was independently proposed by Basov and Prokhorov, by Dicke, and by Schawlow and Townes. In their paper[52] Schawlow and Townes understood immediately that this was a crucial point in laser design and discussed their choice of a Fabry-Perot cavity at length. It is remarkable also that the first lasers and almost all present working lasers use the same basic principle, although a number of different configurations have been introduced. Indeed, the original proposal of using plane parallel mirrors was employed by Maiman and Javan. However, plane mirrors are difficult to align and the silver or aluminium coating for good reflective surfaces, as first used, is also critical.

Immediately after the first laser achievements people began to improve resonator characteristics. The first improvement was the use of a dielectric coating, which decreases losses, improves the quality of the mirror and also allows an initial selection of the wavelength to be amplified in the active material. Javan, in his first He–Ne laser, where a very high reflectivity was needed due to the small gain of the medium, used 13-layer evaporated dielectric films on fused silica plates flat to $\lambda/100$. The layers were alternate films of zinc sulphide and magnesium fluoride which gave a maximum reflectance of 98.9% ± 0.2%, transmittance of 0.3%, and scattering and absorption losses of 0.8%[57, 68].

The two Fabry-Perot plates of the cavity were mounted within the gas chamber (figure 6.10(*b*)) to eliminate the losses that would be caused by unwanted reflections at windows.

Most technical complications occurring in the construction of this first laser could be avoided by simply putting reflectors outside. Rigrod *et al*[69] realised the first laser with external mirrors using the strong 1.1530 μm transition in the He–Ne mixture. The laser beam was allowed to pass through flat windows whose surface normals were placed at the Brewster angle with respect to the beam axis (figure 6.13). This arrangement eliminates Fresnel reflection loss for radiation polarised in the plane of incidence and prevents oscillations in the polarisation normal to this plane.

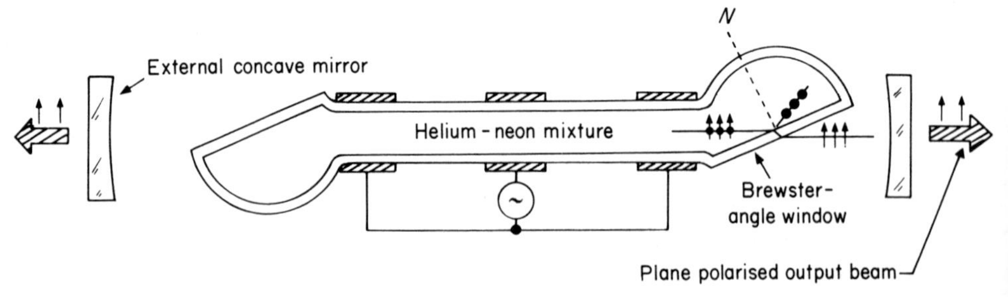

Figure 6.13 The external mirrors confocal cavity used by Rigrod *et al*[69], using Brewster angle windows.

The resultant losses in window transmission come entirely from scattering by imperfections or absorption in the glass and, with reasonable care, may be kept below about 0.5%. The tolerances ($\simeq \pm 3°$) in setting the window at the Brewster angle are quite coarse and easily met. Another innovation in this set-up was the use of spherical mirrors of equal radius with coincident foci (*confocal system*). This system allows rather low-quality windows ($\sim \frac{1}{4}\lambda$) to be used and the criterion for mirror alignment is less severe than the plane-parallel case by about two orders of magnitude in the plate angles.

The confocal geometry had already been suggested by Connes to ameliorate interferometer performance[70]. It was also suggested for laser cavities on the basis of theoretical studies on diffraction losses by Fox and Li[71, 72], Series[73] and Lewis[72]. Considering a confocal resonator and a plane-parallel resonator each of spacing d and of equal Q, the energy distribution in the former is more concentrated on the axis. Therefore the confocal resonator has a smaller effective mode volume. The volume of active material will thus be less for the confocal than for the plane-parallel resonator. Therefore the confocal resonator requires less pump power than the plane-parallel resonator.

The approximate theory of the plane-mirror resonator, as developed by Schawlow and Townes[52], required a more detailed analysis for diffraction losses. The first satisfactory approach was due to Fox and Li[71, 72] who investigated the effects of diffraction on the electromagnetic field in a Fabry–Perot interferometer in free space. Their approach was to consider a propagating wave which is reflected back and forth by two parallel, plane mirrors. They observed that this is equivalent to the case of a transmission medium comprising a series of collinear, identical apertures cut into parallel and equally spaced black partitions of infinite extent. They assumed initially an arbitrary initial field distribution at the first mirror and proceeded to compute the field produced at the second mirror as a result of the first transit. The newly calculated field distribution was then used to compute the field produced at the first mirror as a result of the second transit. The computation was repeated over and over again for subsequent successive transits, and the authors inquired whether, after many transits, the relative field distribution approached a steady state, if there were more than one steady-state solution and what were the losses associated with these solutions. The self-reproducing distributions of phase and relative amplitude over the aperture that they found may be regarded as the proper resonant modes of the cavity.

This calculation, based on Huyghen's principle, was rather cumbersome: in the end a computer was used to give numerical solutions for rectangular-plane, circular-plane, and confocal-spherical or parabolic mirrors.

A parameter which was useful in the calculation was the quantity (Fresnel number) $N = a^2/\lambda L$, defined for circular mirrors of radius a at a distance L from each other. From diffraction theory, when $L \ll a^2/\lambda$, the centre of one mirror is in the near field of the other regarded as an aperture, and the field in the central region of the second mirror may be calculated from the field on the first by means of geometrical optics. When, on the other hand, $L > a^2/\lambda$, then the Fresnel zones appear and geometrical optics are not adequate for the calculations.

A sample of the distribution of amplitude and phase of the dominant mode (TEM_{00}) obtained by Fox and Li is shown in figure 6.14 for a pair of circular-plane mirrors for $N = 2$, 5 and 10. The undulations on the curves are related to the number of Fresnel zones.

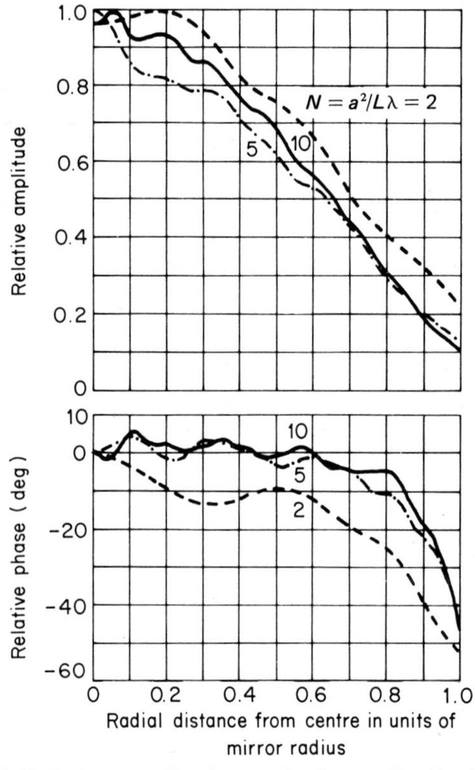

Figure 6.14 Relative amplitude and phase distributions of the dominant (TEM_{00}) mode for circular plane mirrors[72]. Copyright © 1961 American Telephone and Telegraph Company. Reprinted by permission.

The electric-field configurations of modes were also derived and were depicted for square and circular mirrors (figure 6.15).

Using the same method, Boyd *et al*[74, 75] investigated the modes in a resonator formed of two spherical reflectors. They reached many interesting conclusions.

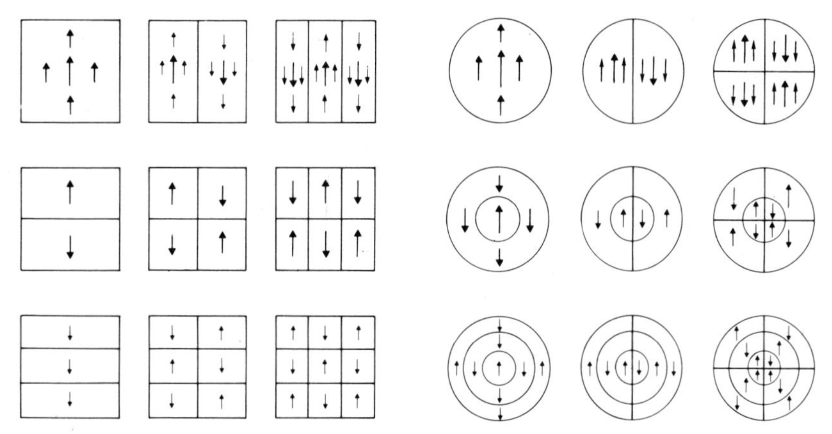

Figure 6.15 Field configuration of normal modes for square and circular mirrors[72]. Copyright © 1961 American Telephone and Telegraph Company. Reprinted by permission.

One was the equality of the resolving power of the Fabry-Perot cavity with its Q, within the small loss approximation; another was the well known Gaussian field distribution of the field in the TEM_{00} mode. They showed that the confocal resonator has several advantages. The diffraction losses are orders of magnitude less than for the plane-parallel resonator. The optical alignment of the two reflectors is not critical. On the basis of an analysis in terms of geometrical optics, Boyd and Kogelnik[75], following a suggestion published later by Fox and Li[76], were able to divide resonators into two classes: stable resonators and unstable resonators. Resonators are stable if

$$0 < [(d/b_1) - 1]\,[(d/b_2) - 1] < 1, \tag{6.15}$$

where the two mirrors of radii b_1 and b_2 are a distance d apart, and unstable whenever this condition is not fulfilled. A bundle of light rays launched in a stable system is periodically refocused as it travels back and forth between the two mirrors, whereas in an unstable system it is dispersed more and more.

A stability diagram for the various resonator geometries was given by Boyd and Kogelnik in the form shown in figure 6.16. According to equation (6.15) the boundary lines between the stable and unstable regions are two straight lines given by

$$d/b_1 = 1 \qquad d/b_2 = 1 \tag{6.16}$$

and a hyperbola which satisfies $d = b_1 + b_2$.

For confocal systems, i.e. systems with coinciding reflector foci, we have $2d = b_1 + b_2$, which may be written,

$$[(d/b_1) - \tfrac{1}{2}]\,[(d/b_2) - \tfrac{1}{2}] = \tfrac{1}{4}. \tag{6.17}$$

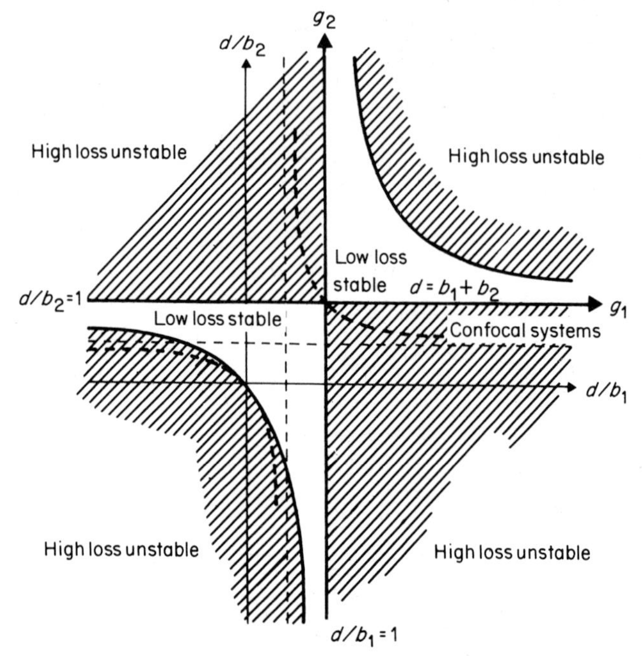

Figure 6.16 Two-dimensional diagram of stable and unstable regions[75].

These systems are represented in figure 6.16 by points on another hyperbola and fall within the high-loss region. Boyd and Kogelnik also showed that the stable and unstable regions can be predicted from the theory of optical modes, and that one must expect relatively high diffraction losses for unstable systems and relatively low diffraction losses for stable systems. This fact was also well demonstrated by Fox and Li[76].

These relations were also shown[77] to derive from the stability of a sequence of equally spaced lenses of equal focal length. Later on the same relation was derived using simple geometrical optics and the matrix method[78].

The results obtained by Fox and Li, and Boyd *et al* were immediately studied and verified by a number of other researchers[79] using various analytical techniques and different kinds of mirrors or configurations. Figure 6.17 shows several possible cavity configurations which were studied.

One of the predictions of the theoretical models concerned the field and theoretical intensity distribution of mode patterns. An approximate description of the transverse (x, y) field distribution and the resonant frequencies of the modes of stable resonators was given in several works[79, 80]. The modes are distinguished by their mode numbers (m, n and q for rectangular geometries, and p, l and q for cylindrical geometries). The mode number q measures the number

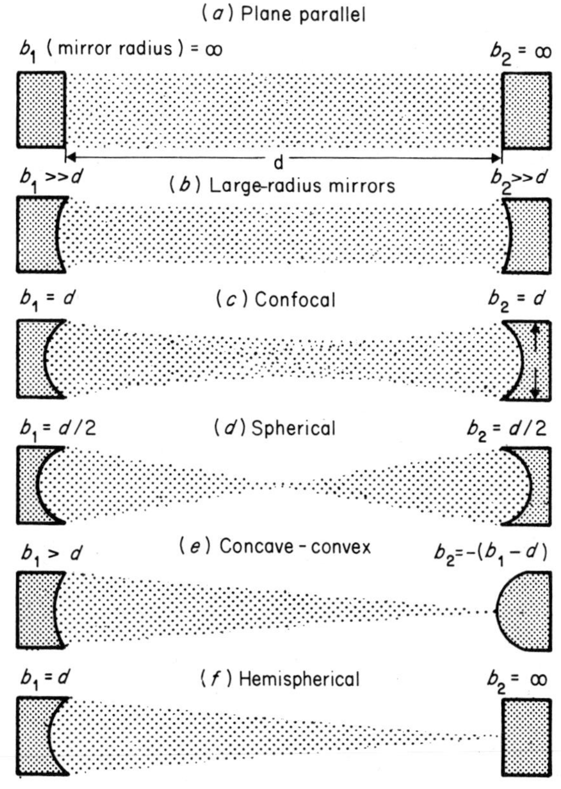

Figure 6.17 Resonator configurations giving uniphase wavefronts (the intra-cavity radiation pattern is shown in light grey). (From A L Bloom: *Spectra Physics Laser Tech. Bull.* 2 August (1963).)

of field zeros of the standing-wave pattern along the z axis. In a rectangular geometry, the transverse mode numbers m, and n measure the field modes in the x and y directions. For rectangular mirrors the field distribution of a TEM_{mnq} mode is given by

$$E(x,y) = E_0 H_m\,[(x/w)\,2^{1/2}]\,H_n\,[(y/w)\,2^{1/2}]\,\exp\,[-\,(x^2+y^2)/w^2], \quad (6.18)$$

where E_0 is a constant amplitude factor, $H_n(x)$ is the Hermite polynomial of nth order and the parameter w is the beam radius or spot size. This measures the beam width of the mode of lowest transverse order ($m = n = 0$), the *fundamental*. According to equation (6.18) the field distribution of this mode is described by a Gaussian profile. A similar formula was obtained for circular mirrors. Formulae for the beam radius were derived by Boyd *et al*[74].

Resonant frequencies were also derived. The frequency spacing ν_0 between the resonances of two modes with the same transverse mode numbers m and n

(or p and l) and neighbouring longitudinal-mode numbers q and $q+1$ was found to be given by

$$\nu_0 = c/2d.$$

For a confocal system[74] the resonant frequencies ν were further given exactly by

$$\nu = \nu_0 \left[(q+1) + \tfrac{1}{2}(m+n+1)\right],$$

and the same for cylindrical symmetry with $(m+n+1)$ replaced by $(2p+l+1)$.

For mirrors of unequal curvature Boyd and Kogelnik[75] derived the approximate expression

$$\nu = \nu_0\{(q+1) + (1/\pi)(m+n+1)\cos^{-1}[(1-d/R_1)(1-d/R_2)]^{1/2}\}$$

where R_1 and R_2 are the radii of curvature of the two mirrors.

The standing-wave pattern of a mode inside the resonator is equivalent to two travelling waves which propagate in opposite directions along the z axis. If a mirror is partially transmitting, the waves incident upon it propagate along the optic axis outside the resonator, thus maintaining the relative field distribution and changing the beam radius, which thus becomes a function of z. Also the phase front $R(z)$ changes. The study of the way these quantities change was started by Yariv and Gordon[81] and others[82] and is well known nowadays under the name of *Gaussian optics*.

The mode-patterns predicted were easily verifiable quantities. Mode patterns of pure rectangular modes of a concave-mirror interferometer were obtained by Kogelnik and Rigrod[83] (figure 6.18) and later by Rigrod[84] for pure circular modes of a concave-plane mirror interferometer (figure 6.19). Mode patterns in ruby were obtained by Evtuhov and Neeland[13].

Attention was also given to unstable resonators by Siegman[85] in 1965, using a simplified geometrical optical analysis. He also pointed out that unstable cavities may be useful in some cases, and gave the following reasons:

(1) unstable resonators can have large mode volume even in very short resonators;

(2) the unstable configuration is readily adapted to adjustable diffraction output coupling; and

(3) unstable resonators have very substantial discrimination against higher-order transverse modes.

The unstable resonator has recently been recognised as being almost ideal for transverse-mode control in TEA lasers, enabling maximum energy extraction in a collimated beam with hard-to-damage reflecting optics[86].

More extensive analyses were done by several groups[87]. Unstable resonators correspond to the shaded regions of figure 6.16. Figure 6.20 shows examples of unstable resonators.

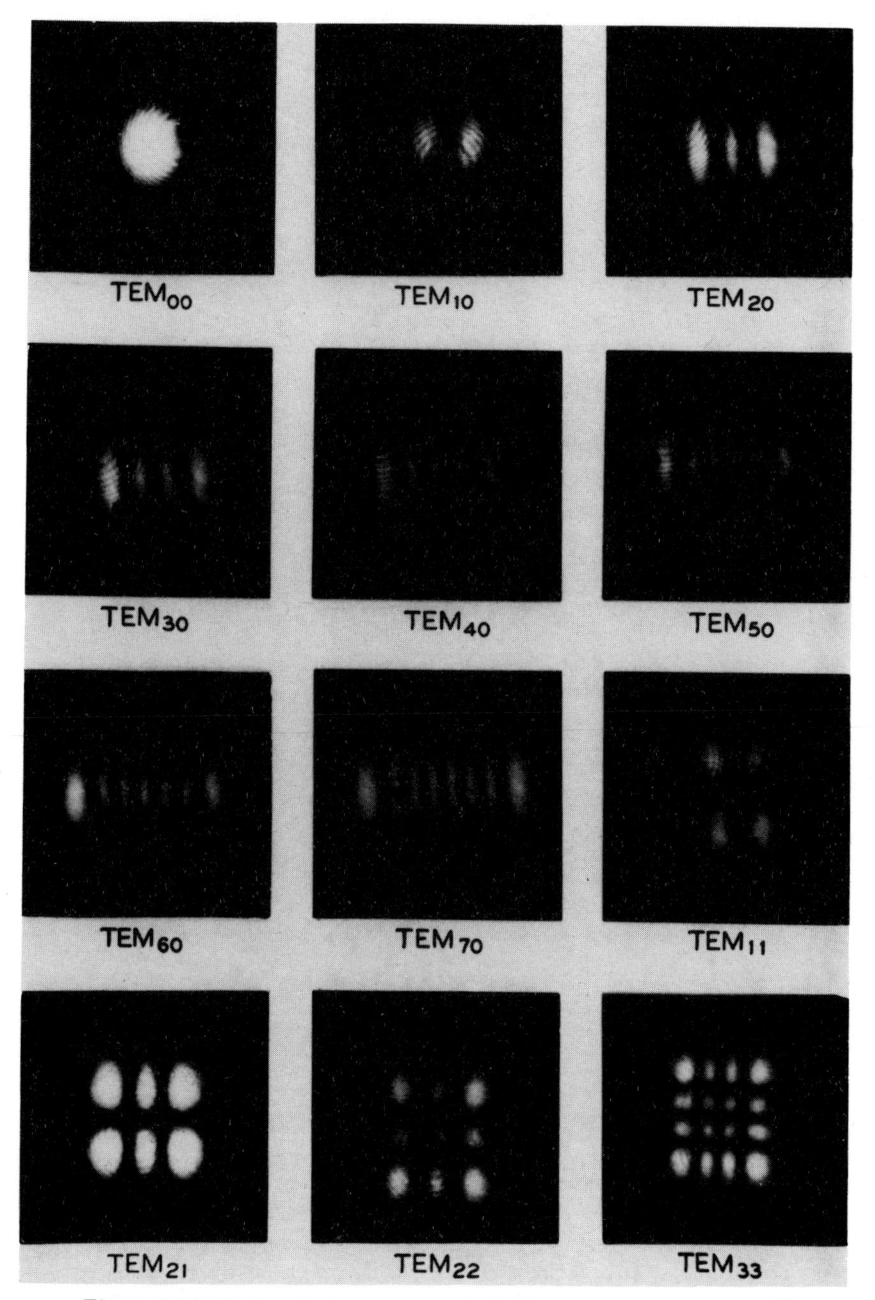

Figure 6.18 Modes in a concave-reflector rectangular cavity[83]. Copyright © 1962 IEEE.

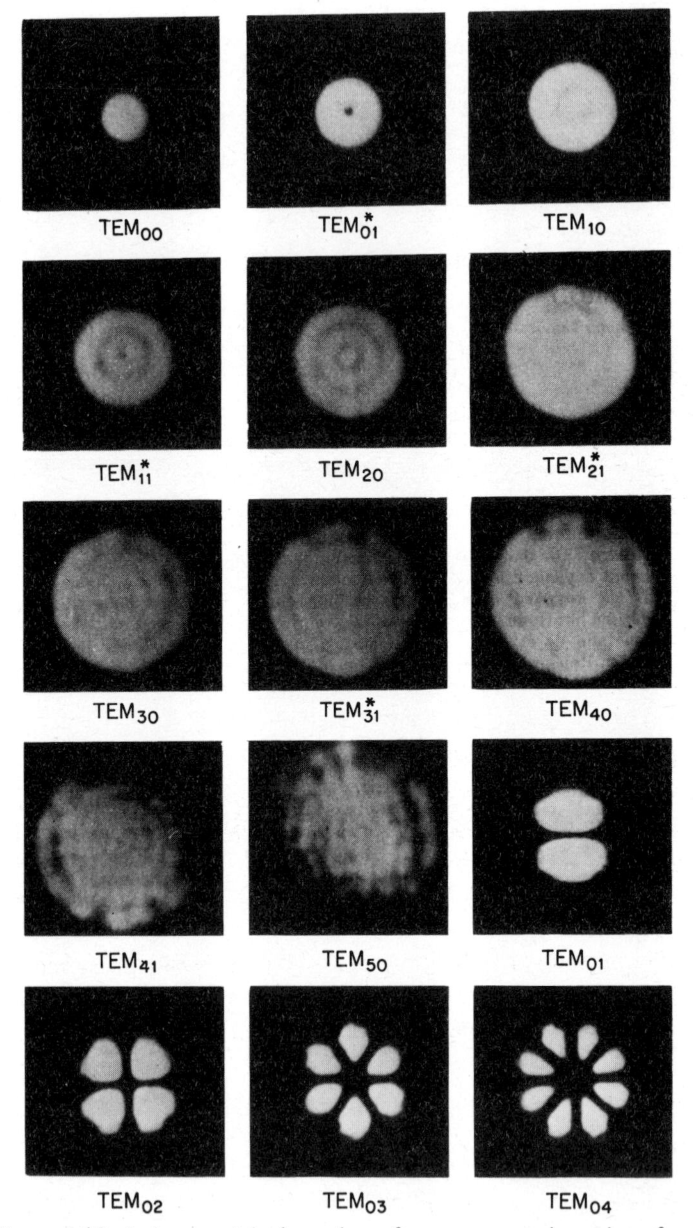

Figure 6.19 Axisymmetrical modes of a concave-mirror interferometer. An asterisk designates two degenerate modes combining in space and phase graduated to form a composite circular-symmetric mode[84].

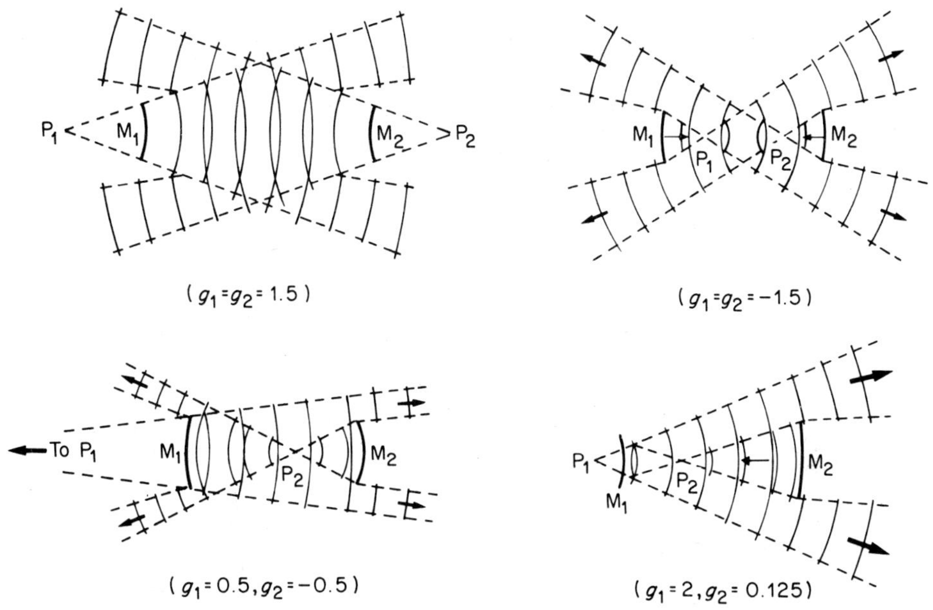

Figure 6.20 Some typical unstable mode patterns illustrating the mode behaviour in different unstable regions of figure 6.16[85]. M_1 and M_2 are the cavity mirrors. Copyright © 1965 IEEE.

Aside from the calculated structures shown in figures 6.17 and 6.20, a number of other configurations were investigated[88]. Probably the most interesting class is formed by resonators made using corner-cube reflectors instead of plane or curved mirrors, as illustrated in figure 6.21. Peck[89] and Murty[90] already used cube-corner prisms in Michelson interferometers to relax alignment tolerances. Peck[91] then showed theoretically that a Fabry–Perot interferometer with cube-corner reflectors replacing one or both flats possesses six eigenpolarisations (i.e. polarisations which are reproduced after traversing the prisms). Gould *et al*[92] considered a cavity consisting of two crossed 90° roof reflectors, replacing the Fabry–Perot mirrors, and showed that it has only two plane eigenpolarisations, still maintaining relaxed alignment tolerances. They also considered the variants roof-reflector versus flat, and the use of entrance faces at or near Brewster's angle. The reflection losses in corner reflectors are virtually eliminated by employing total internal reflection in a prism[93]. A ruby with one end flat (covered with dielectric coating) and the other cut with a 90° dihedral angle like a roof top so as to be effectively 100% reflective was considered by Dayhoff[94]. External glass roof prisms were also used to end ruby rods[95].

Finally, coupling techniques using the phenomenon of frustrated total internal reflection were proposed by several workers[93–6].

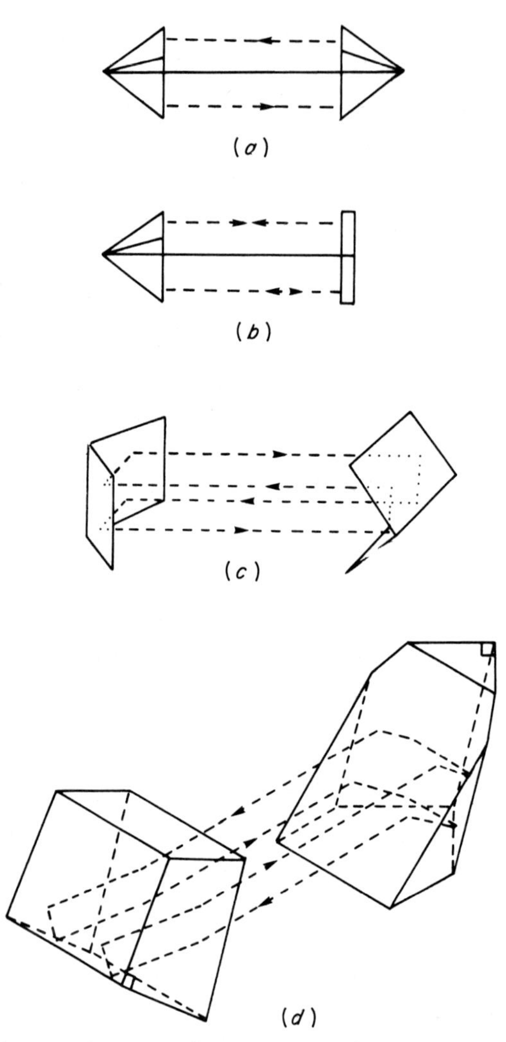

Figure 6.21 Several types of resonators using corner or cube reflectors to form the cavity: (*a*) a pair of corner reflectors cavity[91]; (*b*) a corner and flat reflectors cavity[91]; (*c*) crossed 90° corner reflectors cavity[92]; (*d*) crossed roof prism optical cavity[92]. Reprinted by permission from the Optical Society of America.

Another interesting structure was the one in which the light travelled a closed path, as in a ring. These ring-lasers[97], as they were called, have several interesting properties which can find application, for example in making gyroscopes[98].

A method which had escaped the Fabry–Perot logic was studied by E Snitzer who, in two papers written in 1960[99], considered the use of fibres as dielectric waveguides to provide a resonant structure for an optical maser. Snitzer was an expert in propagation in optical fibres and immediately appreciated the possibility of having a mode selection and a stronger mode coupling in a fibre which was small enough to support only one or two modes. In his analysis he considered a fibre terminated with partially reflecting ends and observed that laser action should be possible at much lower power levels than in an ordinary Fabry–Perot cavity.

At the same time as the study of the mode configuration of the electromagnetic field in an empty cavity, some attempt was made to understand the behaviour of a cavity filled with an active medium.

An analysis by A Kastler[100] pointed out that a Fabry–Perot cavity filled with atoms, even in the case when the system works below the threshold of laser action, shows remarkable properties. In the case of external illumination the local light intensity in the stationary waves inside the interferometer can be much higher than the intensity of the incident light beam. When atoms are emitting, he showed that narrow fringes of very strong intensity are obtained. He also considered two cases of non-uniform distribution of emitting atoms. The first occurred when emitting atoms were located in an atomic beam, showing that Doppler broadening can be suppressed; and the second occurred if atoms had a lamellar distribution between the plates. This last case was a precursor of present distributed-feedback structures[101].

Effects of the laser medium on resonator characteristics, such as mode pulling, were treated by Bennett[102] and Lamb[103].

An understanding of cavity properties enabled substantial improvements to be made in ameliorating the quality of laser output. One step in this direction was mode selection[104].

Another very important improvement came with the understanding that a suitable time control of the Q of the cavity could result in a very large increase in the power output of pulsed lasers. This technique, which is now well known under the name of *Q-switching* or *Q-spoiling*, consists in maintaining a low Q of the cavity and increasing it suddenly some time after pumping has begun, so as to close the cavity when a large increase in population inversion is present. This was introduced by R W Hellwarth[105]. A different method of controlling laser emission arose from the desire to avoid the effect of mode competition during laser emission, which is usually so high that strong amplitude fluctuations and random appearance of different modes are observed. In 1963 S E Harris[106] considered the possibility of applying internal time-varying perturbations to modulate phase and obtain mode control and stability (*mode locking*).

The same result was obtained by introducing modulated losses in the cavity by Hargrove *et al*[107], Gurs and Muller[108] and di Domenico[109], using a 6328 Å He–Ne laser. Di Domenico[109] also gave a linear analysis and discussion.

Locking of modes with phase modulation was later reported in 1964 by Harris and Targ[110] on the same type of laser.

The basic equations which describe the effect of an internal phase or loss perturbation on a laser oscillation were given by several authors[111] and reviewed in a general paper by Harris in 1966[112], where detailed accounts and theory of both FM and AM methods of mode-locking were given.

6.6 Further progress in gaseous lasers

In consequence of these notable advances in the understanding of the properties of an active medium in a resonant cavity, researchers were able to optimise design geometry in their endeavours to construct new lasers.

It could be said with safety that today everything can lase; consequently it would be impossible to trace out the development of all the hundreds of different possible lasers in existence, so we shall confine ourselves here and in the following few sections to a few comments pertaining to some of the most well known lasers, beginning here with gaseous systems.

It is customary to classify lasers obtained in a gas medium according to the following general scheme:

(1) neutral atom lasers;
(2) ion lasers;
(3) molecular lasers;
(4) excimer lasers.

6.6.1 Neutral atom lasers

A typical representative of the neutral atomic gas laser is the He–Ne laser already described.

6.6.2 Ionised gas lasers

Ionised gas lasers use excited ions obtained in an electric discharge. The excitation energy of an ionised atom is much larger than in the case of a neutral atom and therefore emission of these lasers is usually in the visible or near ultraviolet. They also require a much higher current density for the discharge than is required by neutral gases. This is because the lasing level is populated through two or more successive collisions, the first one being to produce the ion and the other one to excite it. As a consequence, these lasers have a high plasma temperature of the order of several thousands of degrees. The first laser action in gaseous ions was obtained by Bell towards the end of 1963. He observed two visible and two infrared transitions between levels of singly ionised mercury[113]. This first ion laser was excited in a pulsed mode by discharging capacitors charged to high voltage into a discharge tube filled with a 500:1 mixture of He:Hg at a pressure of 0.5 Torr. The laser was interesting in that it gave visible wavelength operation ($\lambda = 5678$ Å); high peak powers (up to 40 W); and exhibited high gain (more than 0.8 dB cm^{-1}). The possibility of obtaining cw laser action in an ionic system of this kind also seemed likely, as the length of the laser pulse could be extended by increasing the length of the discharge excitation pulse.

Early in 1964 a report appeared in the magazine *Electronics*[114] of a laser transition observed at a wavelength of 5225 Å in a mercury–argon mixture, and laser action was subsequently reported[115]. The transition was not assigned to either argon or mercury but it may have been the earliest observation of a noble gas ion laser transition, as noted by Bridges[116], who reported the observation of ten laser transitions in the green and blue portions of the visible spectrum between levels of singly ionised argon. These transitions were excited in the pulse mode using either pure argon or mixtures of argon with helium or neon buffer gases. Independently, and almost simultaneously, similar results were reported by Convert *et al*[117] and Bennett *et al*[118]. Bennett *et al*[118] obtained super-radiance and quasi-cw oscillation on several lines and proposed an excitation mechanism through electron impact excitation. Some of the observed transitions exhibited very high gain (several dB m^{-1}) and produced large peak output powers (several hundred watts). Bridges[119] also obtained pulsed ion laser operation in xenon and krypton. Almost immediately afterwards, Gordon *et al*[120] reported continuous laser action in many of the previously reported pulsed transitions in singly ionised argon, krypton and xenon. They used reflecting prisms as described by White[121] to discriminate between the various wavelengths and obtain single wavelength operation. Other authors extended the number of ion laser transitions into the infrared with mercury[122], and with the rare gases into the visible[123], the ultraviolet[124] and the infrared[125].

Gerritsen and Goedertier[126] reported the first laser transition between levels of a doubly ionised atom; mercury.

The discovery of new ion laser transitions had, by the middle of 1965, covered 11 elements with 230 reported transitions[127]. Ten years later, in 1975, ion laser oscillation, had been observed in 32 elements, up to the fifth stage of ionisation in some ion transitions[128].

The most important laser of this category is the Ar^+ laser[116–118, 120]. Due to the high power output attainable in the cw mode of operation and to the convenient wavelength region spanned by the oscillation frequencies (0.45–0.52 μm) the Ar^+ laser soon became one of the most important lasers available. It can be operated in a pure Ar discharge that contains no other gases. The excitation mechanism is discussed in detail in note 128. The pertinent energy level scheme

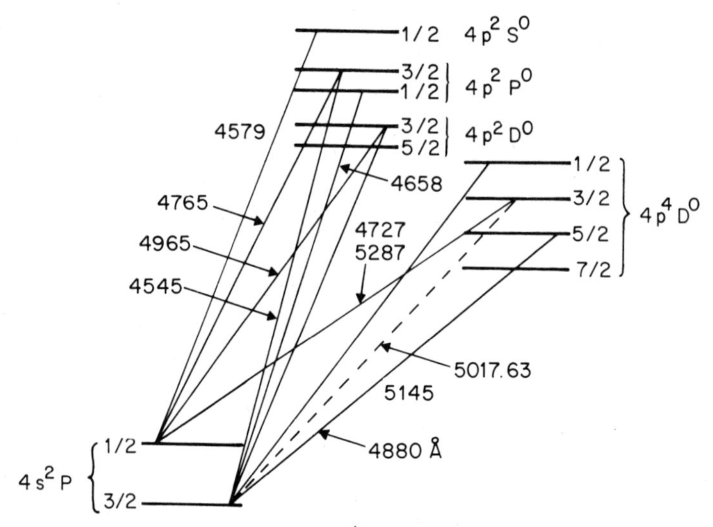

Figure 6.22 Energy levels of Ar^+ and nine laser transitions (from note 116).

was first given by Bridges[116] who also identified most of the observed transitions, and is given in figure 6.22. The more prominent transitions are at 4880 Å and 5145 Å. In the paper the first line was erroneously assigned to the transition $4p^4D^o_{1/2} \rightarrow 4s^2P_{3/2}$ instead of $4p^4D^o_{5/2} \rightarrow 4s^2P_{3/2}$. This line has been corrected in the figure.

6.6.3 Molecular lasers

Molecular lasers use transitions between vibrational and rotational levels of a molecule. Two possible general schemes are feasible. In the first one, transitions between vibrational states of the same electronic level (fundamental) are used. In the second one, vibrational states of different electronic states are used. A typical exponent of the first class is the CO_2 laser, and of the second is the N_2 laser. Both were discovered in 1963.

Legay and Barchewitz in 1963[129] observed strong infrared emission from CO_2 when it was mixed with vibrationally excited nitrogen, and they interpreted this as an indication of collisional energy transfer. Subsequently Legay and Legay-Sommaire[130] suggested on theoretical grounds the possibility of obtaining laser action on the rotation-vibration bands of gases excited by active nitrogen and mentioned specifically the CO_2 10.4 μm transition. At about the same time, Patel *et al*[131] observed laser action in pure CO_2 at 10.6 μm. This discovery came at a time when the search for new laser transitions from ionic and molecular species was near its peak and this particular laser did not attract any more attention than the hundreds of other laser transitions reported at that time. A

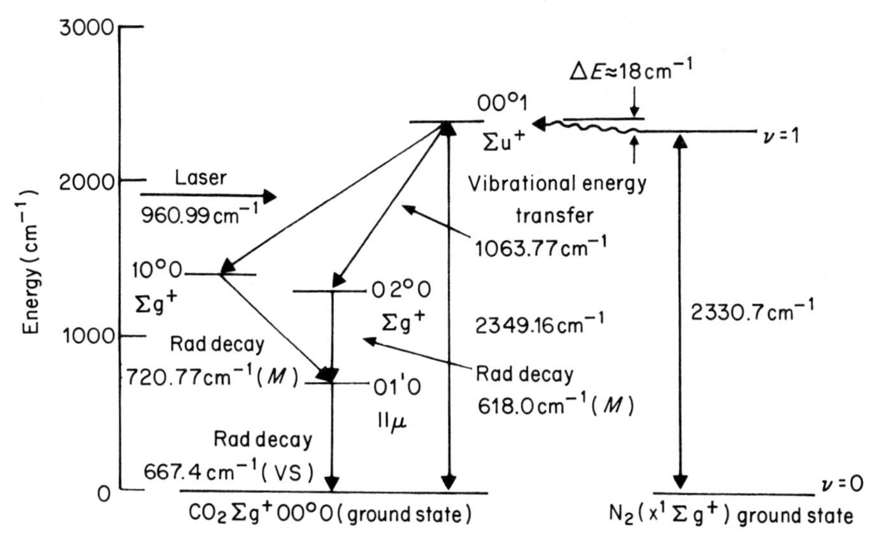

Figure 6.23 Energy-level diagram showing pertinent vibrational levels of CO_2 and N_2[134].

few months later, Patel[132] published a detailed account of his results. Using a 5 m tube he had obtained outputs of 1 mW CW with DC excitation and 10 mW peak with 1 μs excitation pulses. He also presented an interpretation of previously reported results[133]. A short while after, he observed laser action in a CO_2–N_2 mixture[134] and realised that this was a much more efficient mechanism. In this paper he discussed the transfer excitation mechanism and the lasing line which are shown in figure 6.23. The selective excitation of the CO_2 molecule from its ground state to the 00^01 state takes place during a two-body collision involving a CO_2 ground-state molecule and a vibrationally excited N_2 molecule also in its ground electronic state. The first vibrationally excited level ($v = 1$) of N_2 is only 18 cm^{-1} distant in energy from the 00^01 vibrational level of CO_2. The experimental set-up is shown in figure 6.24. It is a continuous-flow system with nitrogen flowing through a high-frequency discharge. The active nitrogen thus formed was mixed with CO_2 in an interaction region between the mirrors of a Fabry–Perot resonator; an output of over 1 mW was observed with a 20 cm long interaction region on the strongest line at 10.6 μm (10.5915 μm). Other lines from 10.5322 to 10.6537 μm were observed, all belonging to the P branch from P(14) to P(26). Further work by Patel[135] and Legay-Sommaire[136] led to the direct excitation of the gas within the laser resonator and the attainment of 12 W from a two-metre tube containing a mixture of CO_2 and air[135]. Howe[137] in the same year investigated the effect of various molecular gases on the laser ouput; he observed oscillation on R-branch lines as well as P-branch lines and noted that the flow rate of the mixture affected the gain. Moeller and Rigden[138]

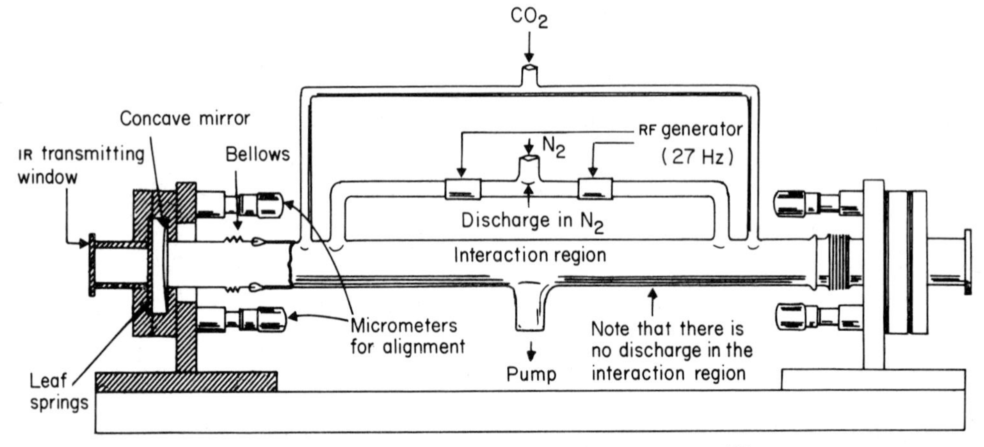

Figure 6.24 The experimental apparatus used by Patel[134] for laser action in N_2–CO_2. (The drawing is not to scale.)

reported 10 W m^{-1} from a sealed tube containing a CO_2–He mixture and also mentioned, with no details, that a flowing mixture of CO_2, N_2 and He gave good results; but the simultaneous report by Patel *et al*[139] gave a detailed account of a water-cooled tube using a flowing CO_2–N_2–He mixture which produced over 50 W m^{-1}. In 1966 *Q*-switching techniques were applied to the CO_2 and N_2O molecular gas laser systems[140]. This paper marks the end of the discovery phase in the history of the CO_2 laser. In 1968 Hill[141] obtained 5 J, 200 kW pulses from CO_2 gas at 10.6 μm at pressures as high as 60 Torr using an ultrahigh-voltage axially pumped discharge. This improvement of nearly three orders of magnitude over previous results was a direct consequence of the use of high-voltage (10^6 V), fast-risetime ($<5\,\mu$s) electrical pulses to excite the CO_2 laser at somewhat elevated pressure (60 Torr). Later on, to meet the requirement of a short discharge time and to lessen the requirement for extremely high applied voltages, pulsed transverse excitation on the CO_2–N_2–He was used[142]. These transversely excited atmospheric-pressure lasers are now called TEA lasers. The technique was immediately applied to several systems[143], and gives high power.

By the mid 1960s it had already been realised that an electron beam might be used both to improve the performance of existing lasers[144] and to extend laser action into the UV portion of the spectrum[145]. The importance of these ideas was not realised until about 1970 or 1971, when the electron-beam-pumped molecular Xe association laser[146] and the electron-beam-controlled discharge atmospheric-pressure CO_2 laser appeared[147].

These developments were quickly followed in 1971 by the first electron-beam initiation of an HF chemical laser[148] and extension of laser action in CO_2 to 25 atm[149] using an electron-beam-controlled discharge. A good and interesting historical survey of the development of pulsed molecular lasers up to 1974 is given by O R Wood II[150].

Finally, although not strictly connected with the CO_2 laser, are gas dynamic lasers. First proposals were by Basov and Oraevskii[151] and other authors[152]. In these lasers the rapid adiabatic expansion of a gas is able to create population inversion between vibrational levels of molecules in a gas mixture which is initially in equilibrium at a very high temperature obtained simply by heating. Three characteristic times are of importance in this case; the relaxation time τ_S of the upper level; the relaxation time τ_L of the lower level and the time needed to reach equilibrium τ_E. If $\tau_E \simeq \tau_L \ll \tau_S$ during expansion, the population of the lower laser level is able to follow the temperature and pressure variations of the gas and so remains in equilibrium with it. However, the population of the upper level is not able to follow the fast variations of conditions of the gas and remains 'frozen' to the equilibrium value at the initial temperature before expansion.

To fulfil these requirements it is necessary to expand the gas through a supersonic nozzle. Gas dynamic lasers usually operate with a mixture of gases (usually CO_2, N_2 and H_2O or He).

N_2 is representative of the class of molecular lasers which use transitions between vibrational states of different electronic states. An initial proposal for laser action in this system was made as early as 1960 by Houtermans[153]. Matthias and Parker[154] were the first to observe laser action in N_2 in 1963, using a peak-pulse voltage and current of about 40 kV and 90 A, respectively, in 2 Torr of nitrogen. They obtained laser action on a number of transitions which have been shown to belong to the first positive system (B $^3\Pi_g$ − A $^3\Sigma^+$) of nitrogen in the range of wavelengths from about 7500 to 12500 Å.

Later Heard[155] obtained laser action in a pulsed nitrogen discharge in the near UV region of the spectrum. The laser transitions extended from about 3000 to 4000 Å and belonged to the second positive system C $^3\Pi_u^+$ − B $^3\Pi_g$ (electronic transitions). The lower laser level here is the upper laser level for the laser action in the near IR. The strongest transition is at 3371.3 Å. In 1965 Leonard[156] observed 200 kW pulses at 3371 Å in N_2 gas at 20 Torr pressure using pulsed transverse excitation, and similar results were also obtained by Gerry[157].

6.6.4 Excimer lasers

The last type of laser to be discussed here is the excimer laser. Excimers are formed by the interaction between two atoms or molecules, one of which is electronically excited

$$A + B^* \rightarrow (AB)^*.$$

The bound molecule AB* may then decay radiatively to the ground state and dissociate

$$(AB)^* \rightarrow A + B + h\nu.$$

The ground state may be either repulsive or sufficiently weakly bound so as to be unstable at normal temperatures. The term *excimer*, first introduced in 1960[158] should properly denote an excited homopolar molecule, e.g. Xe_2^*, whereas for an electronically excited heteropolar complex, e.g. KrF* the term *exciplex* should be used. Excimer lasers are capable of efficient generation of high-power pulses of radiation at ultraviolet and vacuum ultraviolet wavelengths in a spectral region which has no laser transitions. The potential of excimers as laser media was first pointed out by Houtermans[153] in 1960, before the first successful operation of the ruby laser by Maiman. However, the first demonstration of this type of laser action had to wait until 1970, when Basov *et al*[146], following an earlier proposal[159], used liquid xenon pumped with a pulsed high-energy electron beam. At high current densities of the order of 30-60 A cm^{-2}, narrowing of the luminescent line at 1760 Å with a halfwidth reaching 20 Å was observed.

In 1972 the first gaseous excimer using electron-beam excitation of high pressure Xe gas was reported by Koehler *et al*[160] at 173 nm. The first high-power output was subsequently demonstrated by Ault *et al*[161] and Hughes *et al*[162]. These last researchers also obtained laser action in Ar, and laser emission from a krypton excimer was obtained in the same year by Hoff *et al*[163].

6.7 The liquid laser: Dye and chelate lasers

The earliest published works suggesting that organic materials could be used as active media for lasers appear to be those of Brock[164] and Rautian and Sobel'mann[165], who in 1961 proposed that triplet-state phosphorescence could serve as the basis for an organic laser.

In 1964 Stockman, Mallory and Tittel[166] discussed a laser process based upon singlet-state fluorescence and Stockman[167] described early results in the experimental efforts to realise a dye laser using the dye perylene excited by a fast, powerful flashlamp. His efforts, were, however, without success: it is clear in retrospect that the 1 μs flashlamp pulses which he used were too long for the production of laser action in perylene.

The first unambiguously successful attempt to produce stimulated emission from organic molecules was reported in 1966 by Sorokin and co-workers who used a giant ruby laser to excite solutions of the dyes chloroaluminium pthalocianine (CAP)[168] and, a few months later, 3-3′ diethylthiadicarbocyanine (DTTC) iodide in an optical cavity[169].

Similar results were obtained independently, still in the infrared, by Spaeth and Bortfield[170] and by Schäfer *et al*[171] using several cyanine dyes with structures similar to DTTC; and later by Stepanov *et al*[172].

Later still the Sorokin group suggested the use of flashlamp pumping[173] and obtained a working laser of this type[174]; Schmidt and Schafer[175] followed. Visible coherent emission was reported by three different research groups almost simultaneously[176]. In a subsequent ingenious experiment, Soffer and McFarland[177] substituted a diffraction grating for one of the mirrors of a laser-pumped dye laser and obtained efficient spectral narrowing and tunability over a wide spectral range.

One of the most attractive properties of dye lasers is this tunability. In contrast to most other laser media, the emission spectra of fluorescent dyes are broad, permitting the lasing wavelength to be tuned to any chosen value within a fairly broad range. Furthermore, the number of fluorescent dyes is very large and compounds may be selected for emission in any given region of the optical spectrum.

Laser mechanism in dyes can be explained with reference to figure 6.25. State S_0 is the ground state. S_1, S_2, T_1 and T_2 are excited electronic states. States S are singlets, i.e. the spin of the excited electron is antiparallel to the spin of the

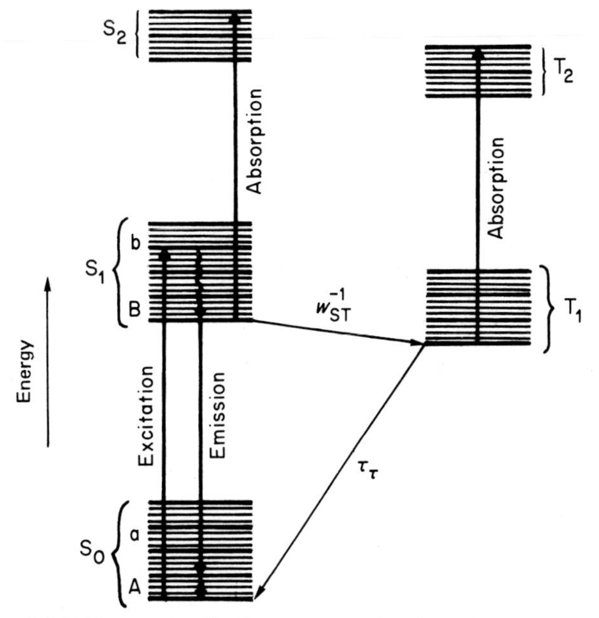

Figure 6.25 Diagram of the energy levels of an organic dye molecule. The heavy horizontal lines represent vibrational states and the lighter lines represent rotational fine structure.

remaining molecule. States T are triplets in which the two spins are parallel. Singlet to triplet, or triplet to singlet transitions are forbidden at first order. Transitions between singlet states or between triplet states give rise to intense absorption of light and to fluorescence. The characteristic colour of organic dyes is due to the absorption from S_0 to S_1. All the states are then split into vibrational and rotational levels as shown in the figure.

Excitation is through the absorption of pumping light which raises the energy of molecules from the fundamental singlet state up to a vibrational level in the first excited singlet state.

This is followed by a very fast non-radiative decay to the bottom of the S_1 levels. Laser emission takes place between this level and a high vibrational level of the fundamental singlet state with a lifetime τ_S. A non-radiative decay from this level to the fundamental state terminates the process.

The vibrational levels of the fundamental state, if sufficiently high in energy, can normally be found unoccupied. It is then possible to have enough population inversion for laser action between the states of the excited singlet and some of the vibrational levels higher than the fundamental state. This condition poses a link between the wavelength of pumping light, the absorption coefficient of the material at that wavelength, and the wavelength at which the laser emission takes place[178].

There is also some probability, W_{ST}, which depends upon the structure of the molecule, that a molecule in the excited singlet level relaxes to a triplet state. Since this transition is forbidden its rate is usually much smaller than τ_S^{-1}. The lifetime τ_T for decay of T_1 to the ground state is relatively long. Therefore the crossing between singlet and triplet is harmful for the laser for two reasons. Firstly, the accumulation of molecules in the triplet state lowers the concentration of molecules available to participate in the laser process. Secondly, and more seriously, the absorption of molecules due to a $T_1 \rightarrow T_2$ transition is very strong. If the wavelength region of this absorption coincides with that of the laser emission, an accumulation of molecules in T_1 will increase the laser losses. For this reason many organic-dye lasers only operate on a pulsed basis.

Another class of liquid lasers consists of organic-rare-earth compounds based upon the fluorescence of rare-earth ions in liquid solvents. The earliest suggestions of the possible use of rare earths in organic compounds for laser action were made in 1962 by Whan and Crosby[179] and Schimitschek and Schwartz[180]. The first laser of this kind, using a class of metallo-organic compound called rare-earth chelate, was reported the following year by Lempicki and Samelson[181] and involved the use of an alcohol solution of the tetrakis chelate, europium benzoylacetonate at 140 K. In these molecules a beta-diketone chelate cage surrounds the rare-earth ion and absorbs light from the pump. This absorbed energy is then transferred to the rare-earth ion which is actually responsible for the lasing process. The following year they obtained operation at room temperature[182].

Chelate lasers have had, however, no great application.

6.8 The chemical laser

In a chemical laser, population inversion and laser output are produced directly by a chemical reaction. Although chemical lasers were discussed theoretically very early in 1960, the first realisation was not made until 1965.

On 8 June 1960 the Canadian chemist J Polanyi, in the section on spectroscopy of a conference of the Royal Society of Canada, proposed an infrared and visible analogue of the maser based on chemical reactions[183]. The point was that chemical reaction products can have negative temperature; so amplification of infrared and visible radiation is possible. He named a device of this type *iraser* or *vaser*. After the conference the paper was sent to *Physical Review Letters*, but the editor rejected it for the reason that lasers were exclusively an engineering topic not of scientific interest. It was subsequently published in the *Journal of Chemical Physics*[184].

One of the reactions proposed in this work, the one with HCl, was later used in the first chemical laser by Kasper and Pimentel[185] in 1965. Polanyi and his staff had already studied excited states by means of infrared chemiluminescence[186], and it came quite naturally to them to think about vibrational states as the source of inverted population, and chemical reactions as one way of creating the population inversion. Lasers based on chemiluminescence were discussed at length during the interval 1960 to 1965[187].

Although chemical lasers were not yet realised, two conferences were dedicated to them: a session on chemical reactions as a possible source of induced radiation in the *1st Symposium on the Elementary Processes of High-energy Chemistry*, Moscow, 18-22 March 1963 and the *1st Conference on Chemical Lasers*, San Diego 9-11 September 1964.

The first chemical laser was built in 1965 by Kasper and Pimentel[185] who demonstrated operation using HCl molecules emitting at 3.7 μm. The excited HCl was generated in a Cl_2–H_2 mixture under pulsed photodissociation of Cl_2 using a flash lamp.

During the next few years the principal workers in the field were chemical kineticists, who found the chemical laser a useful tool for studying the vibrational distribution of reaction products. The photodissociation laser in which a UV flashlamp is used to initiate the chemical laser reaction was used in all of the chemical laser kinetic work during the period 1965-9. Although it is not a chemical laser in the sense that chemical energy is converted to laser emission, it is a laser of considerable importance.

The most highly developed laser of this kind is the iodine laser, for which the typical reactions are

$$h\nu_{\mathrm{UV}} + \mathrm{CF_3I} \rightarrow \mathrm{CF_3} + \mathrm{I}^*(^2\mathrm{P}_{1/2})$$

$$\mathrm{I}^*(^2\mathrm{P}_{1/2}) \rightarrow \mathrm{I}(^2\mathrm{P}_{3/2}) + h\nu_{\mathrm{laser}}.$$

The UV radiation in the band from 2500 to 2900 Å is provided by a flashlamp. Laser emission occurs at 1.315 μm. Typical gases used are CF_3I, CH_3I, C_2H_5I, etc.

In 1967 T Deutsch[188] first used an electric discharge to initiate the chemical laser reaction. He also carried out extensive spectroscopic measurements.

The real growth in the field of the chemical lasers did not occur until 1969, when there were three major developments:

(1) Extensive work with electrically initiated chemical lasers, as reported by Tal'roze, Basov and other Soviet scientists[189].

(2) The first true CW operation of a chemical laser by Spencer[190] who followed previous work by Airey and McKay[191]: in Spencer's work the chemical reagents were mixed rapidly in a supersonic flow stream.

(3) The first purely chemical laser by Cool and Stevens in which bottled gases were mixed directly, without any auxiliary source of reaction initiation, and produced laser output[192].

Later on, pulsed high-energy electron beams[193] came into operation and Zharov[194] showed that laser output energy was about twice the energy put into

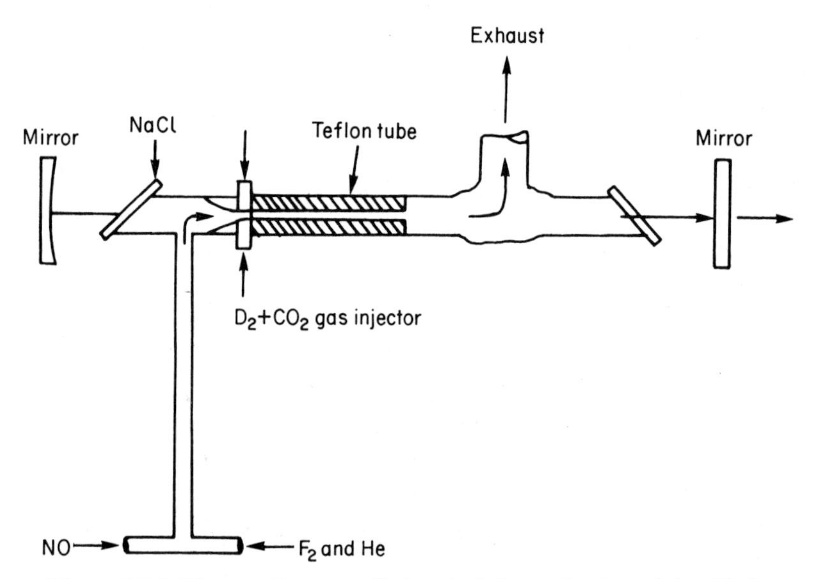

Figure 6.26 The continuous all-chemical laser developed by T A Cool and R R Stephens[192].

the system by the electron beam. All-chemical lasers developed continuously up to 1973 and are discussed in note 195. One of the first realisations is shown in figure 6.26[192].

6.9 The semiconductor laser

Semiconductors were considered at an early stage to offer a good means with which to obtain stimulated emission.

A patent was filed on 22 April 1957 and published later, 20 September 1960[196] by Y Watanabe and J Nishizawa in Japan in which recombination radiation produced by injection of free carriers in a semiconductor was considered. The patent title was *semiconductor maser* and, as a specific example, recombination radiation at about 4 μm in tellurium was considered. Naively, the authors considered the semiconductor in a resonant cavity of the kind used in the microwave region (figure 6.27); but the concept of using an injection of carriers and their recombination radiation was sound. A few months later, in 1958, P Aigrain of France, in a speech at Brussels during an international conference on solid state physics in electronics and telecommunications[197, 198], presented an idea on the extension of maser action to the field of optical frequencies. Unfortunately the paper was not published in the proceedings of the conference.

Bernard and Duraffourg[198] remember he had shown that population inversion between two localised levels is not required in the case of the emission of two bosons, one stimulated and the other thermalised. In semiconductors like germanium and silicon, the valence and conduction band are displaced relative to one another in energy momentum space (figure 6.28). Therefore in the band to band transitions the absorption or emission of a phonon is necessary for conservation of momentum. The longest wavelength emission corresponds to the electron transition from the conduction band minimum to the valence band maximum with the emission of a phonon. The reverse of the above process is a simultaneous absorption of the photon and phonon.

If the sample temperature is sufficiently small and the phonons necessary for the inverse transition are absent, absorption of the long wavelengths under consideration will be small. Therefore, in this case, one can expect that with a slight increase of the carrier concentration compared with the equilibrium concentration, a state with negative temperature relative to the transition under consideration might arise.

In the proceedings of the Brussels conference quoted in note 197, after the summary of the Aigrain paper, at the end of p1766, B Lax observes that the original proposal of an injection semiconductor maser had already been made by John von Neumann in a private communication to John Bardeen in 1954.

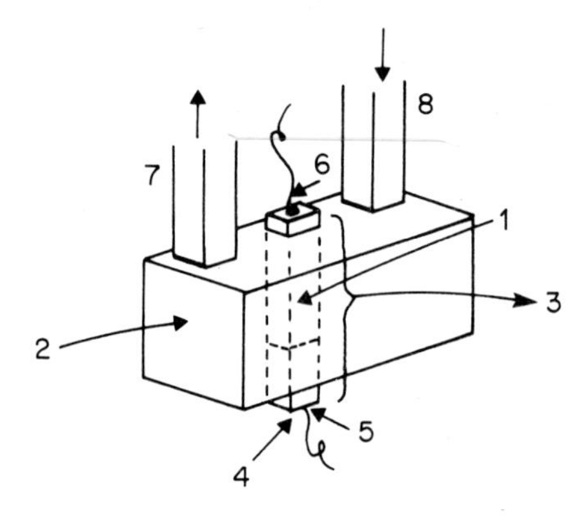

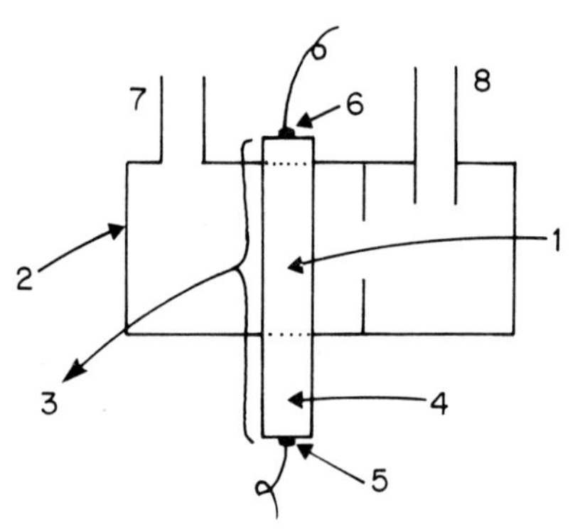

Figure 6.27 Japanese patent for a semiconductor maser[196]. Tellurium is used as a semiconductor (3) in the resonant cavity (2). Holes from region (4) are injected into the semiconductor region (1) and recombine with free electrons, resulting in a radiation of wavelength 4 μm. Electrodes (5) and (6) provide electrical contacts. Output (7) and input (8) terminals are indicated for the amplified radiation.

In the collected works by von Neumann[199] dated 16 September 1953, reviewed by John Bardeen, the following notes on the photon disequilibrium amplification scheme appear:

> The possibility of making a light amplifier by use of stimulated emission in a semiconductor is considered. By various methods, for example by injection of minority carriers from a p–n junction, it is possible to upset the equilibrium concentrations of electrons in the conduction band and holes in the valence band. Recombination of excess carriers may occur primarily by radiation, with an electron dropping from the conduction

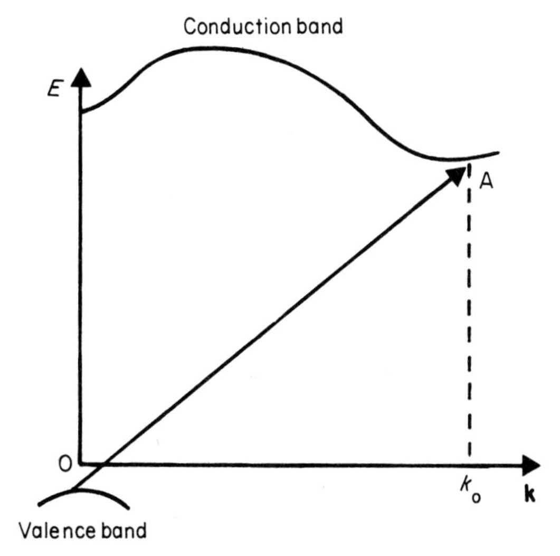

Figure 6.28 Band structure in germanium. If holes are created in the valence band (in O) and electrons in the conduction band (in A) they cannot recombine directly without the help of a phonon of wave vector $\mathbf{k}_0$.

to the valence band and the energy emitted as a photon with an energy slightly greater than the energy gap. The rate of radiation may be enhanced by incident radiation of the same frequency in such a way as to make an amplifier. The basic principle was used later by Townes and by Bloembergen in the MASER (Microwave Amplification by Stimulated Emission of Radiation), although not with recombination radiation in a semiconductor.

Calculations are made for an idealised semiconductor with effective masses in both valence and conduction bands equal to the ordinary electron mass. It is concluded that a very large concentration of excess carriers is required ($\sim 10^{21}$ cm^{-3}). No method is given for obtaining and maintaining large excess carrier concentrations in this concentration range. It is concluded that the large amount of heat released by the radiation might be carried away without requiring an excessive rise in temperature.

Also in 1959 Kromer[200] proposed the use of negative effective mass of carriers to amplify electromagnetic radiation. This proposal was criticised by various authors[201] who showed that a necessary condition for amplification is the formation of states at negative temperature. Zeiger[202] also considered laser action in semiconductors.

In the Soviet Union a group at the Lebedev Institute of Physics (FIAN) of the Academy of Sciences of the USSR headed by Basov had been discussing the possibility of obtaining states with negative temperatures in semiconductors since 1959[203-5]. In the first of these works[203] the use of the impurity ionisation mechanism in a semiconductor at low temperatures using an electric pulse is proposed. After a sufficiently rapid removal of the electric field, and given that the temperature of the crystal lattice is low, it is possible to have negative temperatures due to the difference between the recombination (τ) and slowing-down times of charge carriers in the bands, (t_s), if $t_s \ll \tau$. Later on the slowing-down time t_s was calculated[206] and found to be of the order of 10^{-10} s in Ge and Si, much shorter than the lifetime. This mechanism led, several years later[207], to laser action.

In 1960 a different system was considered in which use was made of the radiation emitted in the electron-hole recombination between bands whose minima are not facing[204]. Semiconductors such as germanium and silicon which present indirect transitions were considered in depth. This mechanism is the same as in the two-boson laser considered by Aigrain.

In 1961 the Soviet group proposed a different method, that eventually proved successful, utilising p-n junctions in degenerate semiconductors[208].

Historically, electroluminescence was observed by Round as early as 1907, when a current was passed through a silicon carbide detector[209]. Similar results were obtained by Lossev in 1923[210], who made a systematic study of the effect and published a series of papers on the subject[211].

The theory of the effect investigated by Lossev was given by Lehovec, Accardo and Jangochian[212], who proposed a model based on the band-structure diagram of a p-n junction. In their model electrons are injected across a forward-biased p-n junction and combine with holes in the p-type region of the junction. In this recombination process the energy lost by the electron is emitted as a photon. In 1961 the group at the Lebedev Institute also considered experimentally excitation of a semiconductor. Excitation with a stong electric field was considered in InSb[213]: only indirect demonstration of the formation of negative temperature states was obtained. Later, a study of negative photoconductivity in some samples of silicon was performed[214].

At the same time, at the *1st* and *2nd Quantum Electronics Conference*, where semiconductor lasers were discussed in depth, B Lax[215] suggested the use of transitions between energy levels generated in strong magnetic levels (Landau levels) or transitions between levels of some impurities to obtain generation in the submillimetre region or in the IR. However, the very short relaxation times of these levels coupled with the presence of absorption transitions as higher levels are approached, created difficulties in generation.

Later, Popov[216] showed that due to the shortening of the lattice relaxation time with current it is practically impossible to obtain a negative temperature state.

The possibility of obtaining stimulated emission in semiconductors by transitions between conduction and valence bands was discussed in a complete and exhaustive manner in 1961 by Bernard and Duraffourg[217]. By using the concept of Fermi quasi-levels they obtained a fundamental relation which must be fulfilled to enable there to be a laser effect in semiconductors. This relation was also derived independently by the Basov group[218].

Bernard and Duraffourg considered that, in the one-particle approximation, an electronic state in a solid may be represented by a Bloch wave defined over the whole crystal. They considered two such states: one in the valence band with a wave vector k_i and energy $E_v(k_i)$, the other in the conduction band with a wave vector k_j and energy $E_c(k_j)$.

If the crystal is not in equilibrium, under certain conditions the occupation probability of any state of the conduction band is given by

$$f_c = [1 + \exp(E(k) - F_c)/kT]^{-1}$$

where F_c is the 'quasi Fermi level' for the electrons of the conduction band. In the same way for holes in the valence band one may introduce a 'quasi Fermi level' F_v. At equilibrium $F_c = F_v = F_0$. These 'quasi Fermi levels' are useful for the description of carriers in a p–n junction.

Let us now consider the case in which a state $E_v(k_i)$ of the valence band is connected to a state $E_c(k_j)$ of the conduction band by a direct radiative transition. Let W_v^c be the probability per unit time of such a process. In a radiation field containing a density of photons $P(\nu)$ of energy $h\nu$ the number N_a of quanta absorbed per unit time is

$$N_a = A W_v^c f_v(k_i) [1 - f_c(k_j)] P(\nu).$$

The number N_e of quanta emitted per unit time by stimulated emission is

$$N_e = A W_c^v [1 - f_v(k_i)] f_c(k_j) P(\nu),$$

where the proportionality coefficient A includes the densities of states of the valence band and of the conduction band. The necessary condition for amplification is now

$$N_e > N_a,$$

or

$$f_c(k_j) [1 - f_v(k_i)] > f_v(k_i) [1 - f_c(k_j)],$$

if it is considered that $W_v^c = W_c^v$. This condition is equivalent to

$$\exp(F_c - F_v)/kT > \exp[E_c(k_j) - E_v(k_i)]/kT.$$

Since

$$E_c(k_j) - E_v(k_i) = h\nu,$$

the above condition simply reduces to

$$F_c - F_v > h\nu. \tag{6.19}$$

In this case a population inversion is not necessary for amplification. This distinct difference from the other kinds of lasers we have considered is due to the continuous distribution of electron levels when compared with the discrete distribution present in atoms or molecules.

Relation (6.19) refers to an energetic condition which allows photons of energy $h\nu$ fulfilling it only to be emitted, not absorbed.

Bernard and Duraffourg also considered some material where condition (6.19) was likely to be fulfilled and in which radiative recombination was large, and suggested among others the III-V compounds GaAs and GaSb.

After the publication of the paper by Bernard and Duraffourg, many groups started active research. In January 1962, Nasledov *et al*[218] in Leningrad reported that the linewidth of the radiation emitted from GaAs diodes at 77 K narrowed slightly at high current densities (1.5×10^3 A cm^{-2}). They suggested that this might be a sign of stimulated emission.

In March 1962 at an American Physical Society meeting, S Mayburgh from the General Telephone and Electronics Laboratory, NY, presented, as a post-deadline paper, *Efficient electroluminescence with GaAs diodes at 77 K*[219], and visited the IBM laboratory, where Dumke[220] had already pointed out the importance of using direct energy gap semiconductors (as in GaAs).

In June, Keyes and Quist[221] announced they had constructed GaAs diodes with an internal quantum efficiency estimated as 85%. Earlier, Pankove[222] had pointed out that a GaAs p-n junction would prove an efficient source of infrared radiation.

In July these results were discussed at the Solid State Device Research Conference and a couple of months later, almost simultaneously, several groups announced laser action in p-n junctions in GaAs. In all cases GaAs, cooled at 77 K, pumped with high-intensity current pulses of a few microseconds duration, was used. The lasers which were then realised were: one by the group at General Electric Research Laboratories, Schenectady, NY, directed by Robert N Hall and announced in a paper received on 24 September[223]; a second one on 4 October by the group at IBM, Yorktown Heights, NY, directed by Marshall I Nathan[224]; and a third on 23 October at Lincoln Laboratories of MIT, directed by T M Quist[225]. At the General Electric Laboratory at Syracuse, Holonyak also succeeded in making a semiconductor laser, as described in a paper submitted on 17 October[226]. These first realisations were all very similar.

Hall's device[223] was a cube of 0.4 mm edge with the junction lying in a horizontal plane through the centre. The front and back faces were polished parallel to each other and perpendicular to the plane of the junction, creating a cavity. Current was applied in the form of pulses of 5–20 μs duration, with the diode polarised in the forward direction and immersed in liquid nitrogen. The threshold for laser action was found at about 8500 A cm^{-2}. Radiation patterns and narrowing of the emission line were studied. Below threshold the spectral width was 125 Å, which constricted suddenly to 15 Å at threshold.

Nathan[224] used a somewhat different system. He started out with GaAs junctions made by diffusing Zn into GaAs doped with Te. These diodes were then banded on to a Au-plated kovar washer and the junction was etched to approximately 1×10^{-4} cm^2. In this case no cavity was made. Threshold for laser emission at 77 K was given in the range 10^4–10^5 A cm^{-2}.

Quist[225] used a mesa structure with an area of 1.4×0.6 mm^2, the short sides being polished optically flat and nearly parallel. At 77 K threshold was found at approximately 10^4 A cm^{-2}: this was decreased by a factor 15 at 4.2 K (figure 6.29). Spectral measurements were also made which confirmed the narrowing of the emission line.

Holonyak and Bevacqua[226] used forward-biased Ga $(As_{1-x}P_x)$ p–n junctions instead. The diodes were rectangular parallelepipedons, or cubes, with two opposite, parallel sides carefully polished so as to give a resonant cavity. Using this

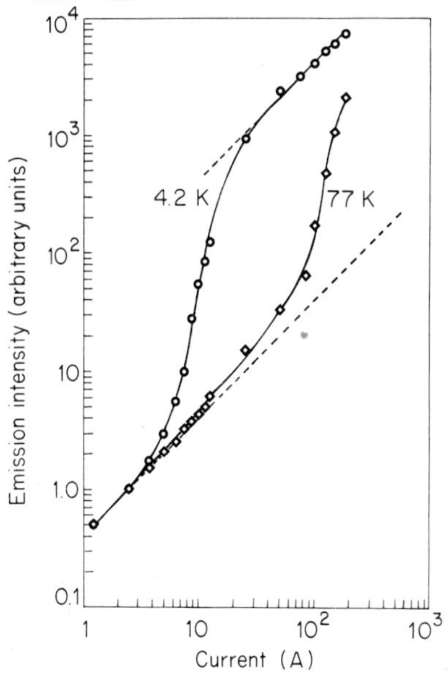

FIgure 6.29 Variation of radiation intensity with current in a GaAs diode at two different temperatures[225].

compound they were able to obtain emission in the region 6000-7000 Å[227] instead of about 8400 Å as obtained using simple GaAs[223, 225].

Almost at the same time as these first results, McWhorter, Zeiger and Lax[228] produced the first theoretical treatment of mode confinement in a p-n junction[229].

In semiconductor lasers transitions take place in a p-n junction between occupied electronic states in the conduction band and empty electronic states in the valence band. A notable difference between these lasers and the other lasers is that here the transitions are between states within which there is a distribution of energy and not between levels of well defined energy.

Figure 6.30(*b*) shows the distribution of the electron states in a doubly degenerate semiconductor. The small dots indicate the energy position of the allowed electron states as a function of the propagation constant k of the electron for both the conduction and valence bands. The situation can exist only under non-thermal equilibrium conditions and can be depicted using two quasi Fermi levels, E_{Fc} and E_{Fv}, in the two bands. If a radiation of frequency ν, such that $E_g < h\nu < E_{Fc} - E_{Fv}$ is propagating through the semiconductor, only downward transitions from occupied conduction-band states to empty valence-band states giving amplification can occur.

The p and n regions in a p-n junction laser are both degenerate. The position of energy bands and electron occupation along a junction, without any bias applied, are shown in figure 6.31.

The Fermi energy has the same value all along the sample (equilibrium condition). If a forward-bias voltage, V_{appl} is applied (figure 6.31(*b*)) the Fermi level in the n region is raised by a quantity eV_{appl} with respect to that in the p region. There now exists a narrow region, the active region of the device, which contains both electrons and holes and is doubly degenerate. Electromagnetic radiation of frequency ν such that

$$E_g/h < \nu < (E_{Fc} - E_{Fv})/h,$$

which is propagating in this region, is amplified.

Many of these first injection lasers were, typically, rectangular parallelepipedons or trapezoids made by cutting chips of material and polishing two parallel ends of the chip. The plane of the p-n junction was perpendicular to the polished ends of the parallelepipedon so as to form a cavity corresponding to a small Fabry-Perot interferometer.

Bond *et al*[230] were the first to make use of cleaving along parallel crystal planes. These lasers, which use a single semiconductor, are usually referred to as *homostructure lasers*. Homostructure injection lasers had very high threshold current densities for lasing at room temperature ($\geqslant$50 000 A cm^{-2}) as a result of which continuous operation was not achieved until 1967[231]. They usually operated at very short pulses ($\leqslant$1 μs) and low duty cycles ($<$0.1%).

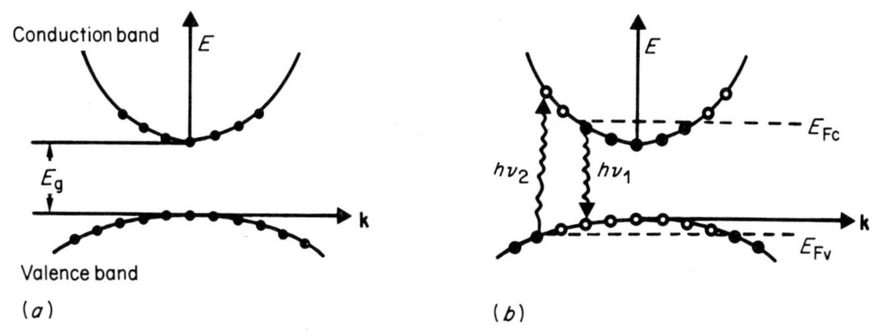

Figure 6.30 Position of the energy levels near the energy gap in a direct semiconductor (such as GaAs) as a function of the electron propagation constant **k**; (*a*) equilibrium situation; (*b*) a doubly degenerate semiconductor.

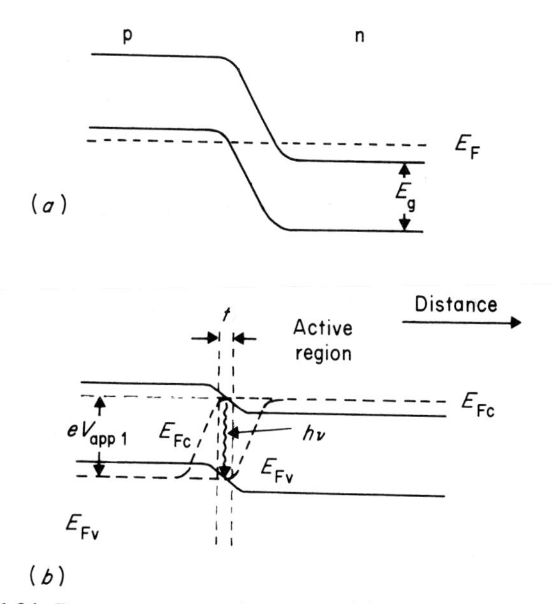

Figure 6.31 Degenerate p–n junction; (*a*) at no applied voltage; (*b*) polarised in the forward direction. Electrons and holes recombine in the junction region.

In 1963, Kroemer[232] suggested the use of heterojunctions, in which a layer of a semiconductor with a relatively narrow energy gap is sandwiched between two layers of a wider energy-gap semiconductor. A similar suggestion was made in the same period by Alverov and Kazarinov and was not even published[233].

Six years elapsed until Hayashi and Panish[234] of Bell Laboratories and Kressel and Nelson[235] at RCA developed the first heterostructure lasers. At the same time Alverov *et al*[236] developed the more complicated multilayer structures which are nowadays called double-heterostructure lasers.

6.10 The free-electron laser

Free-electron lasers arise out of the undulator proposal by Motz in 1951 (Chapter 4), although the first actual proposal that an electron beam could be used to produce coherent light was made much later.

Free-electron lasers are inherently able to produce coherent tunable radiation anywhere in the spectrum between submillimetre and ultraviolet (and perhaps at even shorter wavelengths). Moreover, they are potential sources of very high power, since no damage can occur to the lasing medium, as happens with liquid or solid-state lasers. The basic operating principle comes from the phenomenon of the scattering of photons from electrons in the presence of radiation; that is, stimulated Compton scattering.

In 1933 Kapitza and Dirac[237] predicted stimulated Compton scattering and proposed an experiment to observe the effect with non-relativistic electrons scattered by standing light waves. Much later, Pantel *et al*[238] considered the generation of short-wavelength radiation by stimulated scattering of an electromagnetic pump wave from a relativistic electron beam.

It was, however, only in 1971 that Madey[239], combining all these concepts, analysed the radiation emitted by a relativistic electron beam moving through a periodic, transverse, DC magnetic field. The free relativistic electrons 'see' the virtual quanta representing, in the Weizsacker–Williams approximation, the static, transverse, periodic magnetic field as long-wavelength photons and scatter them into real short-wavelength photons by means of stimulated Compton scattering.

The analogy with the Motz undulator described in Chapter 4 is evident.

In 1972 Madey and co-workers began to build a laser based on these principles. After overcoming innumerable difficulties they finally succeeded in demonstrating stimulated amplification of radiation in 1976[240] and laser oscillation[241] the following year.

The experimental set-up for the demonstration of the amplification of infrared radiation by relativistic free electrons in a constant, spatially periodic, transverse magnetic field is shown in figure 6.32[240].

The electron beam was obtained from the superconducting linear accelerator of the Hansen High Energy Physics Laboratory of Stanford University at an energy of 24 MeV. The periodic magnetic field was generated by a superconducting right-handed double helix with a 3.2 cm period and a length of 5.2 m. The electron beam and an infrared beam from a CO_2 laser ($\lambda = 10.6\ \mu m$) were steered so as to pass through the magnet on the axis.

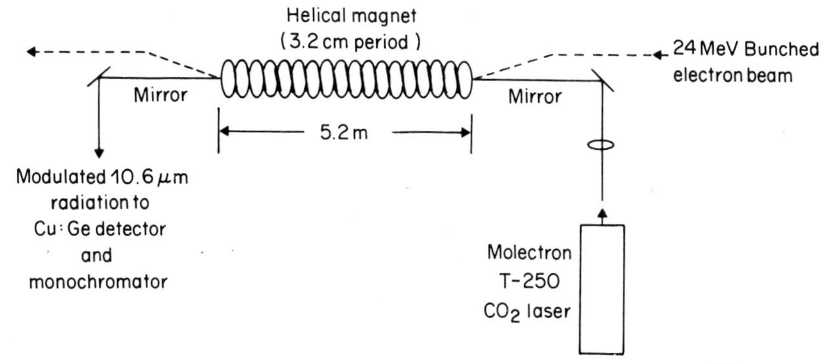

Figure 6.32 Experimental set-up used by L R Elias *et al*[240] to demonstrate amplification of radiation by a free-electron beam. The electron beam was magnetically deflected around the optical components on the axis of the helical magnet.

A gain of 7% per pass was obtained at an electron current of 70 mA. Nearly two years later oscillation at 3.4 μm using a 43 MeV electron beam was demonstrated[241]. Several theoretical works appeared on the subject meanwhile. In 1971, R P Palmer[242] developed a classical picture of how energy would be efficiently transferred between free electromagnetic waves and free, pre-bunched relativistic electrons passing through a static helical magnet, and discussed the possibility of self-induced bunching.

V P Sukhatme and P A Wolff in 1973[243] analysed stimulated Compton scattering using quantum mechanical perturbation theory.

After the first experimental proof of the working principle a number of theoretical papers have also appeared[244].

Notes

1 Operation of a ruby maser at 60 K was reported by C R Ditchfield and P A Forrester, *Phys. Rev. Lett.* **1**, 448 (1958) and, independently, by T H Maiman, *Quantum Electronics,* edited by C H Townes (Columbia University Press: New York, 1960) p234. Operation at 195 K was reported by Maiman, *J. Appl. Phys.* **31**, 222 (1960)

2 T H Maiman, *Phys. Rev. Lett.* **4**, 564 (1960)

3 The spectroscopy of ruby had already been discussed by S Sugano and Y Tanabe, *J. Phys. Soc. Japan* **13**, 880 (1958). The two components R_1 and R_2 of the transition $^4A_2 \rightarrow {}^2E$ were discussed by O Deutschbein, *Ann. Phys. Lpz* **14**, 712 (1932); and *Z. Phys.* **77**, 489 (1932). The notation of levels comes from group theory. When an ion is located in a crystal lattice the electric and magnetic fields prevailing at the site of the ion may exert a profound influence on the energy-level structure of the ion. The number of components into which a multiple level may be split is determined by

group theory; it involves the irreducible representations of the symmetry group applicable to the crystal in question. The split levels are designated by the appropriate symbol of group theory with the spectroscopic symbol of the original level suppressed or abbreviated. See, for example, D S McClure in *Solid State Phys.* **9**, 399 (Academic: New York, 1959). In the case of ruby the ground state of free Cr^{3+} in the crystal is split into three levels denoted by $^4F_1$, $^4F_2$ and $^4A_2$ with multiplicity 12, 12 and 4, respectively. The symbols F_1, F_2 and A_2 refer to the matrix representations of the octahedral group and are not to be related to values of the orbital angular momentum. The upper index 4 is the only reminder that the three levels in question arise from the ground state of the free Cr^{3+} ion which had orbital angular momentum $L = 3$, and spin $S = \frac{3}{2}$. The next lowest group of states of the free Cr^{3+} ion had $L = 4$, $S = \frac{1}{2}$. In the crystal it splits into four sub-levels, which are designated by the symbols $^2A_1$, $^2F_1$, $^2F_2$ and ^2E with multiplicities 2, 6, 6 and 4 respectively. The levels of interest in our discussion are the ones shown in figure 6.1 of the text

4 This circumstance is quoted in B A Lengyel, *Am. J. Phys.* **34**, 903 (1966)

5 T H Maiman, *Nature* **187**, 493 (1960); sentences reproduced by permission Copyright © 1960 McMillan Journals Limited

6 T H Maiman, *British Commun. and Electronics* **7**, 674 (1960)

7 T H Maiman, *Phys. Rev.* **123**, 1145 (1961)

8 T H Maiman, R H Hoskins, I J D'Haenens, C K Asawa and V Evtuhov, *Phys. Rev.* **123**, 1151 (1961)

9 R J Collins, D F Nelson, A L Schawlow, W Bond, C G B Garrett and W Kaiser, *Phys. Rev. Lett.* **5**, 303 (1960)

10 C Kikuchi, J Lambe, G Makhov, R W Terhune, *J. Appl. Phys.* **30**, 1061 (1959)

11 H Statz and G de Mars in *Quantum Electronics* vol 1, edited by C H Townes (Columbia University Press: New York, 1960) p530

12 P Kisliuk and D J Walsh, *Appl. Opt.* **1**, 45 (1962); D F Nelson and R J Collins, *J. Appl. Phys.* **32**, 739 (1961); M Hercher, *Appl. Opt.* **1**, 665 (1962); J I Masters and G B Parrent, *Proc. IRE* **50**, 230 (1962); M D Galanin, A M Leontovich and Z A Chizhikova, *Sov. Phys.-JETP* **43**, 347 (1962); D A Berkley and G J Wolga, *Phys. Rev. Lett.* **9**, 479 (1962); M S Lipsett and L Mandel, *Nature* **199**, 553 (1963); M S Lipsett and L Mandel, *Proc. Third Int. Conf. Quantum Electronics* (1963) vol 3, p 1271; I D Abella and C H Townes, *Nature* **192**, 957 (1961); G R Hanes and B P Stoicheff, *Nature* **195**, 587 (1962); T P Hughes, *Nature* **195**, 325 (1962); M Ciftan, A Krutchkoff and S Koosekanani, *Proc. IRE* **50**, 84 (1962); J Borie and A Orzag *Comptes Rendus* **255**, 874 (1962)

13 V Evtuhov and J K Neeland, *Appl. Opt.* **1**, 517 (1962); these observations were reported in a more complete way in *Quantum Electronics* edited by P Grivet and N Bloembergen (Columbia University Press: New York, 1964) vol 3, p1405 and *IEEE J. Quantum Electron.* **QE-1**, 7 (1965)

14 A E Siegman, *Microwave phototubes and light demodulators, in Northeast Electronics Research and Engineering Meeting (NEREM) Record* (Boston, Mass, 1961); B J McMurtry and A E Siegman, *Appl. Opt.* **1**, 51 (1962)

15 R C Duncan jr, Z J Kiss and J P Wittke, *J. Appl. Phys.* **33**, 2568 (1962)

16 M Ciftan, A Krutchkoff and S Koozekanani, *Proc. IRE* **50**, 84 (1962)

17 A L Schawlow in *Quantum Electronics* vol 1, edited by C H Townes (Columbia University Press: New York, 1960) p553
18 N lines in ruby were studied by several researchers (C E Mendenhall and R W Wood, *Phil. Mag.* **30**, 316 (1915); O Deutschbein, *Ann. Phys. Lpz* **14**, 712 (1932); ibid. **14**, 729 (1932); ibid. **20**, 828 (1932)). Deutschbein labelled them as N for Nebenlinien. A L Schawlow, D L Wood and A M Clogston (*Phys. Rev. Lett.* **3**, 271 (1959)) had shown that they disappear at low concentrations of Cr, but appear strongly in emission at concentrations of the order of 0.5% (red ruby). The energy-level structure of Cr in the red ruby is different from that of pink ruby because of the exchange interaction between the strongly magnetic Cr ions. Fluorescence in red ruby thus occurs at 7009 (N_2) and 7041 (N_1) Å
19 A L Schawlow and G E Devlin, *Phys. Rev. Lett.* **6**, 96 (1961)
20 I Weider and L R Sarles, *Phys. Rev. Lett.* **6**, 95 (1961)
21 F J McClung, S E Schwarz and F J Meyers, *Phys. Rev.* **124**, 3139 (1962)
22 M Hercher (*Appl. Opt.* **1**, 665 (1962)) studied the effects of optical path length variations with a Twyman–Green interferometer. Also Evtuhov and Neeland[13] used this interferometer to study the emission of ruby. A deterioration of the near-field pattern with decrease in optical quality was observed by C M Stickley, *Appl. Opt.* **2**, 855 (1963) and J W Growe, at a meeting of the American Optical Society, Spring 1963, studied the effect of scattering centres on laser-mode formation
23 R L Hutcheson, *IRE Int. Conv. Record* **10**, 147 (1962)
24 *Linde Crystal Products Bulletin* F-2330 (Linde Co: East Chicago, Ill, 1965) see R D Olt, *Appl. Opt.* **1**, 25 (1962)
25 The first ruby lasers were pumped with the rod surrounded by a helical xenon flash lamp, possibly with a reflector over the outside. A right-elliptic cylinder with a lamp along one focus and the laser crystal along the other was used by M Ciftan, C F Luck, C G Shafer and H Statz (*Proc. IRE* **49**, 960 (1961)). A modification with several intersecting ellipses with a common focus containing the crystal was later employed by C Bowness, D Missio and T Rogala (*Proc. IRE* **50**, 1704 (1962)). Rougher and simpler arrangements used the rod placed adjacent to one or more linear flash lamps and surrounded by aluminium foil, magnesium oxide or other highly reflecting material (P A Miles and H E Edgerton, *J. Appl. Phys.* **32**, 740 (1961)). Other dispositions were used by P H Keck, J J Redman, C E White and R E Dekinder jr (*Appl. Opt.* **2**, 827 (1963)), D F Nelson and W S Boyle (*Appl. Opt.* **1**, 181 (1962)) and D Roers (*Appl. Opt.* **3**, 259 (1964)). The efficiency of laser cavities was calculated by O Svelto, *Appl. Opt.* **1**, 745 (1962)
26 D F Nelson and W S Boyle, *Appl. Opt.* **1**, 181 (1962)
27 P P Sorokin and M J Stevenson, *Phys. Rev. Lett.* **5**, 557 (1960)
28 P P Sorokin and M J Stevenson, *Adv. Quantum Electron.* edited by J R Singer (Columbia University Press: New York, 1961) p65
29 See for example: P Pringsheim, *Fluorescence and Phosphorescence* (Wiley: New York, 1949); D S McClure, *Solid State Phys.* **9**, 400 (1959); W A Runciman, *Rep. Prog. Phys.* **21**, 30 (1958). The first spectroscopic studies of $CaF_2{:}U^{3+}$ were those of L N Galkin and P P Feofilov (*Dokl. Akad. Nauk USSR* **114**, 745 (1957) (in Russian) Transl. in *Sov. Phys.*

Doklady **2**, 255 (1957)) and L H Galkin and P P Feofilov (*Opt. Spectrosc.* **1**, 492 (1959)

30 G D Boyd, R J Collins, S P S Porto, A Yariv and W A Hargreaves, *Phys. Rev. Lett.* **8**, 269 (1962); S P S Porto and A Yariv, *J. Appl. Phys.* **33**, 1620 (1962)

31 This result was already known, see B Bleaney, P M Lewellyn and D A Jones, *Proc. Phys. Soc.* B **69**, 858 (1956)

32 S P S Porto and A Yariv, *Quantum Electronics* edited by P Grivet and N Bloembergen (Dunod: Paris, 1964), p717

33 H A Bostick and J R O'Connor, *Proc. IRE* **50**, 219 (1962)

34 P A Miles, *Discussion* in *Advances in Quantum Electronics* edited by J R Singer (Columbia University Press: New York, 1961) p76

35 P P Sorokin and M J Stevenson, *IBM J. Res. Develop.* **5**, 56 (1961). A US patent on a four-level laser based on this laser was assigned to P P Sorokin on 1/11/1966 no. 3.229.306

36 L F Johnson and K Nassau, *Proc. IRE* **49**, 1704 (1961). Also this laser was patented by the authors US patent no. 3.225.306 issued 21 December 1965

37 L F Johnson, G D Boyd, K Nassau and R R Soden, *Proc. IRE* **50**, 213 (1962) and *Phys. Rev.* **126**, 1406 (1962)

38 P H Keck, J J Redman, C E White and D E Bowen, *Appl. Opt.* **2**, 833 (1963)

39 L F Johnson, G D Boyd and K Nassau, *Proc. IRE* **50**, 87 (1962) and *Proc. IRE* **50**, 86 (1962)

40 L F Johnson, *J. Appl. Phys.* **33**, 756 (1962); L F Johnson and R R Soden, *J. Appl. Phys.* **33**, 757 (1962): a résumé of these researches is in note 42

41 B A Lengyel, *Introduction to Laser Physics* (Wiley: New York, 1966) p113

42 L F Johnson, *J. Appl. Phys.* **34**, 897 (1963)

43 G H Dieke and H M Crosswhite, *Appl. Opt.* **2**, 675 (1963)

44 J E Geusic, H M Marcos and L G Van Uitert, *Appl. Phys. Lett.* **4**, 182 (1964)

45 The main reasons for the success of Nd:YAG lasers together with their further developments can be found in H G Danielmeyer's paper in *Lasers* vol 4, edited by A K Levine and A J De Maria (Marcel Dekker: New York, 1976) p1

46 E Snitzer, *Phys. Rev. Lett.* **7**, 444 (1961)

47 E Snitzer, *Quantum Electronics*, edited by P Grivet and N Bloembergen (Dunod: Paris, 1964) p999

48 E Snitzer, *Phys. Rev. Lett.* **7**, 444 (1961); C G Young, *Appl. Phys. Lett.* **2**, 151 (1963)

49 H W Etzel, H W Gandy and R J Ginther, *Appl. Opt.* **1**, 534 (1962); E Snitzer and R Woodcock, *Appl. Phys. Lett.* **6**, 45 (1965); H W Gandy, R J Ginther and J F Weller, *Phys. Lett.* **16**, 3266 (1965); —— *Phys. Lett.* **11**, 213 (1964); H W Gandy and R J Ginther, *Proc. IRE* **50**, 2113 (1962); —— *Appl. Phys. Lett.* **1**, 25 (1962)

50 See note 1 of Chapter 5

51 A Ferkhman and S Frish, *Phys. Z. Sowjet.* **9**, 466 (1936), reported by N G Basov and O N Krokhin, *Appl. Opt.* **1**, 213 (1962)

52 A L Schawlow and C H Townes, *Phys. Rev.* **112**, 1940 (1958)

53 J H Sanders, *Phys. Rev. Lett.* **3**, 86 (1959)

54 A Javan, *Phys. Rev. Lett.* **3**, 87 (1959)

55 W E Lamb and R C Retherford, *Phys. Rev.* **79**, 549 (1950)

56 A Javan, in *Quantum Electronics* edited by C H Townes (Columbia University Press: New York, 1960) p 564

57 A Javan, W R Bennett jr and D R Herriott, *Phys. Rev. Lett.* **6**, 106 (1961). W R Bennett and A Javan had a US patent for this laser no. 3.149.290 issued 15 September 1964

58 The level notation in the figure is the Russell–Sanders notation for helium, and the Paschen notation for neon. In the Russell–Sanders coupling scheme the number of the total angular momentum is indicated with a capital letter as follows: S for $L = 0$, P for $L = 1$, D for $L = 2$, F for $L = 3$, G for $L = 4$, H for $L = 5$, I for $L = 6$, K for $L = 7$. If necessary, the value of the total electron spin added to the angular momentum is indicated as a small subscript on the right. As a superscript at the left is the multiplicity of the state or the number of possible values of the vector sum of an orbital angular momentum and a spin angular momentum. So, for example, the ground state of helium which is $1s^2$ with the two electron spins antiparallel to each other, is indicated as $1\,^1S$. The first excited state 1s2s is $2\,^3S$ if the spins are parallel and $2\,^1S$ if the spins are antiparallel. Coupling of the outer-electron of an excited noble gas to the core electrons does not follow the Russell–Sanders rules. Paschen notation is a system of shorthand symbols. The ground state of Neon is $2p^6$ which in Paschen notation is simply indicated as 1p. The excited configurations are $2p^5 3s$ (in Paschen notation 1s), $2p^5 3p$ (in Paschen notation 2p), $2p^5 4s$ (in Paschen notation 2s), $2p^5 5s$ (3s), and so on

59 D R Herriott, *J. Opt. Soc. Am.* **52**, 31 (1962); —— in *Advances in Quantum Electronics* edited by J R Singer (Columbia University Press: New York, 1961) p44

60 W R Bennett jr, *Phys. Rev.* **126**, 580 (1961); W R Watson and T G Polanyi, *J. Appl. Phys.* **34**, 708 (1963)

61 A D White and J D Rigden, *Proc. IRE* **50**, 1697 (1962)

62 A L Bloom, W E Bell and R C Rempel, *Appl. Opt.* **2**, 317 (1963)

63 N G Basov and O N Krokhin, *JETP* **39**, 1777 (1960) (in Russian); *Sov. Phys.-JETP* **12**, 1240 (1961); N G Basov, O N Krokhin and Yu M Popov, *Usp. Fiz. Nauk* **72**, 1961 (1960) (in Russian)

64 N G Basov and O N Krokhin, *Appl. Opt.* **1**, 213 (1962)

65 V K Ablekov, M S Pesin and I L Fabelinski, *Zh. Eksp. Teor. Fiz.* **39**, 892 (1960) (in Russian); *Sov. Phys.-JETP* **12**, 618 (1960)

66 V K Ablekov, *Sov. Phys.-JETP* **42**, 736 (1962)

67 P Rabinowitz, S Jacobs and G Gould, *Appl. Opt.* **1**, 513 (1962); W R Bennett jr, W L Faust, R A MacFarlane and C K N Patel, *Phys. Rev. Lett.* **8**, 470 (1962); A D White and J D Rigden, *Proc. IRE* **80**, 1697 (1972); C K N Patel, W R Bennett jr, W L Faust and R A McFarlane, *Phys. Rev. Lett.* **9**, 102 (1962); R A McFarlane, C K N Patel, W R Bennett jr and W L Faust, *Proc. IRE* **50**, 2111 (1962)

68 The first use of dielectric mirrors for a ruby laser seems to have been by McClung *et al*[21] to select emission on the R_2 line against the one on the

R_1; Sorokin and Stevenson[27] had already used them in CaF_2:U^{s+} and CaF_2:Sm^{2+}

69 W W Rigrod, H Kogelnik, D J Brangaccio and D R Herriott, *J. Appl. Phys.* **33**, 743 (1962)

70 P Connes, *Revue d'Optique* **35**, 37 (1956); *J. Phys. Radium* **19**, 262 (1958)

71 A G Fox and T Li, *Proc. IRE* **48**, 1904 (1960)

72 A G Fox and T Li, *Bell Syst. Tech. J.* **40**, 453 (1961); see also *Advances in Quantum Electronics* edited by J Singer (Columbia University Press: New York, 1961) p308

73 Quoted in G D Boyd and J P Gordon, *Bell Syst. Tech. J.* **40**, 489 (1961)

74 G D Boyd and J P Gordon, *Bell Syst. Tech. J.* **40**, 489 (1961); this paper immediately followed the one of Fox and Li (note 72). See also G D Boyd in *Advances in Quantum Electronics* edited by J Singer (Columbia University Press: New York, 1961) p318; G D Boyd in *Quantum Electronics* edited by P Grivet and N Bloembergen (Columbia University Press: New York, 1964) p1193

75 G D Boyd and H Kogelnik, *Bell Syst. Tech. J.* **41**, 1347 (1962)

76 A G Fox and T Li, *Proc. IRE* **51**, 80 (1963)

77 G Goubau and F Schwering, *IRE Trans. Antennas and Propagation* (May 1961) **AP-9**, p256; J R Pierce, *Proc. Nat. Acad. Sci.* **47**, 1808 (1961); F Schwering, *Arch. Electr. Uber.* **15**, 555 (1961)

78 M Bertolotti, *Nuovo Cim.* **26**, 401 (1962)

79 G Toraldo di Francia in *Quantum Electronics and Coherent Light* edited by P A Miles (Academic: New York, 1964) p53; C L Tang, *Appl. Opt.* **1**, 768 (1962); J M Burch in *Quantum Electronics* edited by P Grivet and N Bloembergen (Columbia University Press: New York, 1964) p1187; M Pouthier, ibid. p1253; A G Fox and T Li, ibid. p1263; J Kotik and M C Newstein, *J. Appl. Phys.* **32**, 178 (1962); R F Soohoo, *Proc. IEEE* **51**, 70 (1963); S R Barone, *J. Appl. Phys.* **34**, 831 (1963); J B Beyer and E H Scheibe, *IRE Trans. Antennas Propagation* **10**, 349 (1962); A G Fox and T Li, *Proc. IEEE* **51**, 80 (1963); W Culshaw, *IRE Trans. MTT* **10**, 331 (1962); H Kogelnik and T Li, *Proc. IEEE* **54**, 1312 (1966); L A Vainshtein, *JETP* **17**, 709 (1963); *Sov. Phys.-Tech. Phys.* **9**, 157 (1964); J P Gordon and H Kogelnik, *Bell Syst. Tech. J.* **43**, 2873 (1964); C Y She and H Heffner, *Appl. Opt.* **3**, 703 (1964)

80 L A Vainshtein, *JETP* **18**, 471 (1964); *Sov. Phys.-Tech. Phys.* **9**, 166 (1964); P O Clark, *Proc. IEEE* **53**, 36 (1965); *J. Appl. Phys.* **36**, 66 (1965); W A Specht, *J. Appl. Phys.* **36**, 1306 (1965)

81 W Yariv and J P Gordon, *Proc. IEEE* **51**, 4 (1963)

82 For example N Kambe, *Proc. IEEE* **52**, 327 (1964); K Miyamoto, *J. Opt. Soc. Am.* **54**, 989 (1964); S A Collins, *Appl. Opt.* **3**, 1263 (1964); S A Collins and D T M Davis, *Appl. Opt.* **3**, 1314 (1964); T Li, *Appl. Opt.* **3**, 1315 (1964); J P Gordon, *Bell Syst. Tech. J.* **43**, 1826 (1964)

83 H Kogelnik and W W Rigrod, *Proc. IRE* **50**, 220 (1962)

84 W W Rigrod, *Appl. Phys. Lett.* **2**, 51 (1963)

85 A E Siegman, *Proc. IEEE* **53**, 277 (1965)

86 A E Siegman, *Laser Focus* **7**, 42 (May 1971)

87 A E Siegman and R W Arrathoon, *IEEE J. Quant. Electron.* **QE-3**, 156

(1967), and many others. For a general review see A E Siegman, *Appl. Opt.* **13**, 353 (1974)

88 For example, R J Collins and J Giordmaine in *Quantum Electronics III* vol 2, edited by P Grivet and N Bloembergen (Columbia University Press: New York, 1964) p1239, considered a ruby crystal in the shape of a rectangular parallelepipedon. P A Kleiman and P P Kisliuk, *Bell Syst. Tech. J.* **41**, 453 (1962) considered a double mirror ended interferometer which allowed discrimination against unwanted modes

89 E R Peck, *J. Opt. Soc. Am.* **38**, 66 (1948); ibid. **38**, 1015 (1948); ibid. **47**, 250 (1957)

90 M V R K Murty, *J. Opt. Soc. Am.* **50**, 7 (1960); ibid **50**, 83 (1960)

91 E R Peck, *J. Opt. Soc. Am.* **52**, 253 (1962)

92 G Gould, S Jacobs, P Rabinowitz and T Shultz, *Appl. Opt.* **1**, 533 (1962); see also P Rabinowitz, S F Jacobos, T Schultz and G Gould, *J. Opt. Soc. Am.* **52**, 452 (1962)

93 L Bergstein, W Kahn and C Shulman, *Proc. IRE* **50**, 1833 (1962)

94 E S Dayhoff, *Proc. 10th Colloquium Spectroscopicum International* 1962, p421

95 M Bertolotti, L Muzii and D Sette, *Nuovo Cim.* **26**, 401 (1962)

96 H A Daw, *J. Opt. Soc. Am.* **53**, 915 (1963); I N Court and F K Willisen, *Appl. Opt.* **3**, 719 (1964); D F Holshouser, *Quantum Electronics* vol 3, edited by P Grivet and N Bloembergen (Columbia University Press: New York, 1964) p1453, suggested the use of conical ends for total internal reflection

97 A H Rosenthal, *J. Opt. Soc. Am.* **52**, 1143 (1962)

98 W M Macek and D T M Davis jr, *Appl. Phys. Lett.* **2**, 67 (1963)

99 E Snitzer, *J. Opt. Soc. Am.* **51**, 491 (1961); *J. Appl. Phys.* **32**, 36 (1961)

100 A Kastler, *Appl. Opt.* **1**, 17 (1962)

101 The major distinction between a distributed feedback laser and a conventional laser is that the distributed feedback laser does not use cavity mirrors. Instead, feedback is provided via Bragg scattering from spatially periodic perturbation of the optical parameters of the active medium. Such parameters can be the refractive index of the laser medium as in the first realisation by H Kogelnik and C V Shank, *Appl. Phys. Lett.* **18**, 152 (1971) and later used by R L Fork, K R German, and E A Chandross, *Appl. Phys. Lett.* **20**, 139 (1972); I P Kaminow, H P Weber and E A Chandross, *Appl. Phys. Lett.* **18**, 497 (1971), the optical gain or both of these together as in C V Shank, J E Bjorkholm and H Kogelnik, *Appl. Phys. Lett.* **18**, 395 (1971); J E Bjorkholm and C V Shank, *Appl. Phys. Lett.* **20**, 306 (1972); J E Bjorkholm and C V Shank, *IEEE J. Quant. Electron.* **QE-8**, 833 (1972); S Chandra, N Takenchi and S R Hartmann, *Appl. Phys. Lett.* **21**, 144 (1972); or the waveguide cross section in the case of thin film lasers (see P Zory, *Appl. Phys. Lett.* **22**, 125 (1973)

102 W R Bennett jr, *Appl. Opt. Suppl. on Optical Masers* **1**, 24 (1962)

103 W E Lamb jr, *Phys. Rev.* **134**, A1429 (1964)

104 In note (88) Kleiman and Kisliuk's paper is one of the first examples of mode selection. This field grew enormously since the first experiments. A review of first efforts is given by P W Smith in *Lasers* vol 4, edited by A K

Levine and A J De Maria (Marcel Dekker: New York, 1976) p74

105 R W Hellwarth, *Advances in Quantum Electronics* edited by J R Singer (Columbia University Press: New York, 1961) p334. In this paper the proposal was made with no experimental demonstration. F J McClung and R W Hellwarth, *J. Appl. Phys.* **33**, 828 (1962), *Proc. IEE* **51**, 46 (1963); R W Hellwarth, *Phys. Rev. Lett.* **6**, 9 (1961); the use of a rotating disc was made by R J Collins and P Kisliuk, *J. Appl. Phys.* **33**, 2009 (1962); see also N G Basov, V S Zuev and P G Krjukov, *Appl. Opt.* **1**, 1 (1962); A J De Maria, R Gagosz and G Barnard, *J. Appl. Phys.* **34**, 453 (1963) considered use of an ultrasonic shutter. Use of a photosensitive liquid was discussed by J I Masters, J Ward and E Hartouni, *Rev. Sci. Instrum.* **34**, 365 (1963)

106 The proposal was made by S E Harris in October 1963 at Stanford University under Contract AF 33 (657) – 11144 as quoted in note 110

107 L E Hargrove, R L Fork and M A Pollack, *Appl. Phys. Lett.* **5**, 4 (1964)

108 K Gurs and R Muller, in *Proc. Symp. on Optical Masers; New York 1963* (Polytechnic Press: Brooklyn, New York) p243; K Gurs, in *Quantum Electronics III* edited by P Grivet and N Bloembergen (Columbia University Press: New York, 1964) p1113

109 M J Di Domenico jr, *J. Appl. Phys.* **35**, 2870 (1964)

110 S E Harris and R Targ, *Appl. Phys. Lett.* **5**, 202 (1964)

111 E I Gordon and J D Rigden, *Bell Syst. Tech. J.* **42**, 155 (1963); A Yariv, *J. Appl. Phys.* **36**, 388 (1965); *IEEE J. Quantum Electron.* **QE-2**, 30 (1966); S E Harris and O P McDuff, *Appl. Phys. Lett.* **5**, 205 (1964); *IEEE J. Quantum. Electron.* **QE-1**, 245 (1965); M H Crowell, *IEEE J. Quantum Electron.* **QE-1**, 12 (1965)

112 S E Harris, *Proc. IEEE* **54**, 1401 (1966)

113 W E Bell, *Appl. Phys. Lett.* **4**, 34 (1964)

114 *Electronics* p17, 24 January 1964

115 H G Heard, G Makhov and J Peterson, *Proc. IEEE* **52**, 414 (1964)

116 W B Bridges, *Appl. Phys. Lett.* **4**, 128 (1964)

117 G Convert, M Armand and P Martinot-Lagarde, *Comptes Rendus* **258**, 4467 (1964); **258**, 3259 (1964)

118 W R Bennett jr, J W Knutson jr, G N Mercer and N J L Detch, *Appl. Phys. Lett.* **4**, 180 (1964)

119 W B Bridges, *Proc. IEEE* **52**, 843 (1964)

120 E I Gordon and E P Labuda, *Bell Syst. Tech. J.* **43**, 1827 (1964); F I Gordon, E F Labuda and W B Bridges, *Appl. Phys. Lett.* **4**, 178 (1964)

121 A D White, *Appl. Opt.* **3**, 431 (1964)

122 A L Bloom, W E Bell and F O Lopez, *Phys. Rev.* **135**, A578 (1964)

123 P Laures, L Dana and C Frapard, *Comptes Rendus* **258**, 6363 (1964); **259**, 745 (1964)

124 L Dana and P Laures, *Proc. IEEE* **53**, 78 (1965)

125 F A Horrigan, S H Koozekanani and R A Paananen, *Appl. Phys. Lett.* **6**, 41 (1965)

126 H J Gerritsen and P V Goedertier, *J. Appl. Phys.* **35**, 3060 (1964)

127 W B Bridges and A N Chester, *IEEE J. Quantum Electron.* **QE-1**, 66 (1965)

128 C C Davis and T A King in *Advances in Quantum Electronics* vol 3 (Academic: New York, 1975) p169

129 F Legay and P Barchewitz, *Comptes Rendus* **256**, 5304 (1963)
130 F Legay and N Legay-Sommaire, *Comptes Rendus* **259**, 99 (1964)
131 C K N Patel, W L Faust and R A McFarlane, *Bull. Am. Phys. Soc.* **9**, 500 (1964)
132 C K N Patel, *Phys. Rev.* **136**, A1187 (1964)
133 C K N Patel, *Phys. Rev. Lett.* **12**, 588 (1964)
134 C K N Patel, *Phys. Rev. Lett.* **13**, 617 (1964)
135 C K N Patel, *Appl. Phys. Lett.* **7**, 15 (1965)
136 N Legay-Sommaire, L Henry and F Legay, *Comptes Rendus* **260**, 3339 (1965)
137 J A Howe, *Appl. Phys. Lett.* **7**, 21 (1965)
138 G Moller and J D Rigden, *Appl. Phys. Lett.* **7**, 274 (1965)
139 C K N Patel, P K Tien and J H McFee, *Appl. Phys. Lett.* **7**, 290 (1965)
140 M A Kovacs, G W Flynn and A Javan, *Appl. Phys. Lett.* **8**, 62 (1966). A somewhat unusual technique was proposed in the same year by T J Bridges (*Appl. Phys. Lett.* **9**, 174 (1966)) by using a moving mirror technique
141 A E Hill, *Appl. Phys. Lett.* **12**, 324 (1968)
142 R Dumanchin and J Rocca-Serra, *Comptes Rendus* **269**, 916 (1969); A J Beaulieu, *Appl. Phys. Lett.* **16**, 504 (1970)
143 O R Wood, E G Burkhardt, M A Pollack and T J Bridges, *Appl. Phys. Lett.* **18**, 261 (1971)
144 W R Bennett jr, *Appl. Opt.* **2**, 3 (1965); P K Tien, C MacNair and H L Hodges, *Phys. Rev. Lett.* **12**, 30 (1964); Yu V Tkach, Ya B Fainberg, L I Bulotin, Ya Ya Bessarab and N P Gadetsuii, *JETP Lett.* **6**, 371 (1967)
145 N G Basov, *IEEE J. Quantum Electron.* **QE-2**, 354 (1966); A G Molchanov, I A Polacktrova and Yu M Popov, *Sov. Phys.-Solid State* **9**, 2655 (1968)
146 N G Basov, V A Danilychev, Yu M Popov and D D Khodkevich, *JETP Lett.* **12**, 329 (1970)
147 N G Basov, *JETP Lett.* **14**, 285 (1971); *Sov. J. Quantum Electron.* **1**, 306 (1971); J D Daugherty, E R Pugh and D H Douglas-Hamilton, *Bull. Am. Phys. Soc.* **17**, 399 (1972); C A Fenstermacher, M J Nutter, W T Leland and K Boyer, *Appl. Phys. Lett.* **20**, 56 (1972); C A Fenstermacher, M J Nutter, J P Rink and K Boyer, *Bull. Am. Phys. Soc.* **16**, 42 (1971)
148 D W Gregg, *Chem. Phys. Lett.* **8**, 609 (1971); Y I Pan, C E Turner jr, K J Pettipiece, *Chem. Phys. Lett.* **10**, 577 (1971)
149 N G Basov *et al*, *JETP Lett.* **14**, 285 (1971)
150 O R Wood II, *Proc. IEEE* **62**, 355 (1974)
151 N G Basov and A N Oraevskii, *JETP* **17**, 1171 (1963)
152 N G Basov *et al*, *Sov. Phys.-Tech. Phys.* **13**, 1630 (1969); E T Gerry, *Appl. Phys. Lett.* **76** (1965)
153 F G Houtermans, *Helv. Phys. Acta* **33**, 933 (1960)
154 L E S Matthias and J T Parker, *Appl. Phys. Lett.* **3**, 16 (1963)
155 H G Heard, *Nature* **200**, 667 (1963); *Bull. Am. Phys. Soc.* **9**, 65 (1964)
156 D A Leonard, *Appl. Phys. Lett.* **7**, 4 (1965)
157 E T Gerry, *Appl. Phys. Lett.* **7**, 6 (1965)
158 B Stevens and E Hutton, *Nature* **186**, 1045 (1960)
159 N G Basov, *IEEE J. Quantum Electron.* **QE-2**, 354 (1966)
160 H A Koehler, H A Ferderber, D L Redhead and P J Ebert, *Appl. Phys. Lett.* **21**, 198 (1972)

161 E R Ault *et al*, *IEEE J. Quantum Electron.* **9**, 1031 (1972)
162 W M Hughes, J Shannon, A Kolb, E Ault and M Bhaumik, *Appl. Phys. Lett.* **23**, 385 (1973)
163 P W Hoff, J C Swingle and C K Rhodes, *Appl. Phys. Lett.* **23**, 245 (1973)
164 E G Brock, P Czavinsky, H Hormats, H C Nedderman, D Stirpe and F Unterleitner, *J. Chem. Phys.* **35**, 759 (1961)
165 S G Rautian and I I Sobel'mann, *Opt. Spectrosc.* **10**, 134 (1961); *Opt. Spectrosc.* **10**, 65 (1961)
166 D L Stockman, W R Mallory and F K Tittel, *Proc. IEEE* **52**, 318 (1964)
167 D L Stockman, *Proc. 1964 ONR Conf. on Organic Lasers*
168 P P Sorokin and J R Lankard, *IBM J. Res. Develop.* **10**, 162 (1966)
169 P P Sorokin, W H Culver, E C Hammond and J R Lankard, *IBM J. Res. Develop.* **10**, 401 (1966)
170 M L Spaeth and D P Bortfield, *Appl. Phys. Lett.* **9**, 179 (1966)
171 F P Schaefer, W Schmidt and J Volze, *Appl. Phys. Lett.* **9**, 306 (1966)
172 B I Stepanov, A N Rubinov and V A Mostovsikov, *JETP Lett.* **5**, 117 (1967)
173 P P Sorokin, J R Lankard, E C Hammond and V L Moruzzi, *IBM J. Res. Develop.* **11**, 130 (1967)
174 P P Sorokin and J R Lankard, *IBM J. Res. Develop.* **11**, 148 (1967)
175 W Schmidt and F P Schafer, *Z. Naturf.* **229**, 1563 (1967)
176 F P Schafer, W Schmidt and K Marth, *Phys. Lett.* **24A**, 280 (1967); P P Sorokin, J R Lankard, E C Hammond and V L Moruzzi, *IBM J. Res. Develop.* **11**, 130 (1967); B B McFarland, *Appl. Phys. Lett.* **10**, 208 (1967)
177 B H Soffer and B B McFarland, *Appl. Phys. Lett.* **10**, 266 (1967)
178 B I Stepanov and A N Rubinov, *Sov. Phys.-Usp.* **11**, 304 (1968)
179 R E Whan and G A Crosby, *J. Mol. Spectrosc.* **8**, 315 (1962)
180 E J Schimitschek and E G K Schwartz, *Nature* **196**, 832 (1962)
181 A Lempicki and H Samelson, *Phys. Lett.* **4**, 133 (1963); *Proc. Symp. on Optical Masers* (Polytechnic Press: Brooklyn, New York, 1963) p347
182 H Samelson, A Lempicki, C Brecher and V Brophy, *Appl. Phys. Lett.* **5**, 173 (1964)
183 J C Polanyi, *Proc. R. Soc. V Canada* **54**, 25 (1960)
184 J C Polanyi, *J. Chem. Phys.* **34**, 347 (1961)
185 J V V Kasper and G C Pimentel, *Phys. Rev. Lett.* **14**, 352 (1965)
186 See for example J C Polanyi, *J. Chem. Phys.* **31**, 1338 (1959); J K Cashion and J C Polanyi, *Proc. R. Soc.* **A 258**, 529 (1960); J K Cashion and J C Polanyi, *J. Chem. Phys.* **29**, 455 (1958)
187 N G Basov, A N Oraevskii, *JETP* **44**, 1742 (1963) (in Russian); G Karl and J C Polanyi, *Discuss. Faraday Soc.* **33**, 93 (1962); A N Oraevskii, *JETP* **45**, 177 (1963) (in Russian); V L Tal'roze, *Kinetica i Katalize* **1**, 11 (1964) (in Russian); A N Oraevskii, *JETP* **48**, 1150 (1965) (in Russian); R A Young, *J. Chem. Phys.* **40**, 1848 (1964); and finally, *Appl. Opt. Suppl. 2: Chemical Lasers* 1965 which gives an impressive look at the research in the field up to the moment of the first realisation
188 T F Deutsch, *Appl. Phys. Lett.* **10**, 234 (1967)
189 N G Basov, E P Markin, A I Nikitin and A N Oraevskii in *Proc. Symp. Chemical Lasers, St. Louis* (1969); O M Batovskii, G K Vasil'ev, E F Makarov and V L Tal'roze, *JETP Lett.* **9**, 200 (1969); N G Basov, L V

Kulakov, E P Markin, A I Nikitin and A N Oraevskii, *JETP Lett.* **9**, 375 (1969); see also *Int. Symp. Chemical Lasers, Moscow, USSR* 2–4 September 1969

190 D J Spencer, T A Jacobs, J H Mirels and R W F Gross, *Int. J. Chem. Kinetics* **1**, 493 (1969), Addendum ibid. **2**, 337 (1970); D J Spencer, H Mirels, T A Jacobs and R W F Gross, *Appl. Phys. Lett.* **16**, 235 (1970)

191 J R Airey and S F McKay, *Appl. Phys. Lett.* **15**, 401 (1969). They obtained CW operation for 1.8 ms, using a shock-tube geometry

192 T A Cool and R R Stephens, *Appl. Phys. Lett.* **16**, 55 (1970); *J. Chem. Phys.* **51**, 5175 (1969); T A Cool, R R Stephens and T J Falk, *Int. J. Chem. Kinetics* **1**, 495 (1969)

193 Y I Pan, C E Turner jr and K J Pettipiece, *Chem. Phys. Lett.* **10**, 577 (1971)

194 V F Zharov, V K Malinovskii, Yu S Neganov, G M Chumak, *JETP Lett.* **16**, 154 (1972)

195 T A Cool, *IEEE J. Quantum Electron.* **QE-9**, 72 (1973)

196 Japanese Patent no. 273217, 20 September 1960, Yasushi Watanabe, Sendai and Jun-ichi Nishizawa, Sendai, Appl. no. 32-9899 filed 22 April 1957: *Semiconductor maser*

197 P Aigrain, *Congrès Internationel sur le Physique de l'Etat Solide et ses Applications à l'Electronique et aux Télécommunications* Bruxelles 1958, quoted by P Aigrain in *Quantum Electronics* edited by P Grivet and N Bloembergen (Dunod: Paris, 1964) p1761

198 M G A Bernard and G Duraffourg in *Quantum Electronics* edited by P Grivet and N Bloembergen (Dunod: Paris, 1964) p1849

199 J von Neumann, *Collected Works* vol 5 (Pergamon: London, 1963) p420

200 N Kromer, *Proc. IRE* **47**, 397 (1959)

201 N G Basov, O N Krokhin and Yu M Popov, *Zh. Eksp. Teor. Fiz.* **38**, 1001 (1960); D C Matthis and M J Stevenson, *Phys. Rev. Lett.* **3**, 18 (1959); P Kaus, *Phys. Rev. Lett.* **3**, 20 (1959)

202 H J Zeiger in *Quarterly Progress Report on Solid State Research, Lincoln Laboratory, MIT* (15 October 1959) pp41–3 AD 231991

203 N G Basov, B M Vul and Yu M Popov, *Zh. Eksp. Teor. Fiz.* **37**, 587 (1959) (in Russian); *Sov. Phys.-JETP* **10**, 416 (1960)

204 N G Basov, O N Krokhin and Yu M Popov, *Zh. Eksp. Teor. Fiz.* **39**, 1001 (1960); *Zh. Eksp. Teor. Fiz.* **39**, 1486 (1960) (in Russian); *Sov. Phys.-JETP* **12**, 1033 (1961); *Zh. Eksp. Teor. Fiz.* **40**, 1203 (1961) (in Russian); *Sov. Phys.-JETP* **13**, 845 (1961); in *Advances in Quantum Electronics* edited by J R Singer (Columbia University Press: New York, 1961) p496

205 N G Basov, *Quantum Electronics* edited by P Grivet and N Bloembergen (Dunod: Paris, 1964) p1769

206 O N Krokhin and Yu M Popov, *Sov. Phys.-JETP* **38**, 1589 (1960)

207 The mechanism led to laser action in CdS by F H Nicoll, *Appl. Phys. Lett.* **23**, 465 (1973) and later by the Lebedev group in several semiconductors, under the name of *semiconductor streamer lasers*; see N G Basov *et al*, *JETP* **70**, 1751 (1976); N G Basov, A G Molchanov, A S Nasibov, A Z Obidin, A N Pechenov and Yu M Popov, *IEEE J. Quantum Electron.* **QE-13**, 699 (1977); *JETP* **19**, 336 (1972); *JETP Lett.* **19**, 336 (1974); *IEEE J. Quantum Electron.* **QE-10**, 794 (1974)

208 N G Basov, O N Krokhin and Yu M Popov, *JETP* **10**, 1879 (1961)
209 T Round, *Electrical World* 309 (1907)
210 O V Lossev, *Telegraphia i Telefonia* **18**, 61 (1923)
211 O V Lossev, *Phil. Mag.* **6**, 1024 (1928)
212 K Lehovec, C A Accardo and E Jamgochian, *Phys. Rev.* **83**, 603 (1951)
213 N G Basov, B D Osipov and A N Khvoshchev, *Zh. Eksp. Teor. Fiz.* **40**, 1882 (1961); B D Osipov and A N Khvoshchev, *Zh. Eksp. Teor. Fiz.* **43**, 1179 (1962)
214 N G Basov, O N Krokhin, L M Lisitsyn, E P Markin and B D Osipov, *Zh. Eksp. Teor. Fiz.* **41**, 988 (1961). A historical perspective is presented in Yu M Popov *Proc. PN Lebedev Physics Institute*, vol 31 (Consultant Bureau: New York, 1968)
215 B Lax in *Quantum Electronics* edited by C H Townes (Columbia University Press: New York, 1960) p428, and *Advances in Quantum Electronics* edited by J Singer (Columbia University Press: New York, 1961) p465
216 Yu M Popov, *Fiz. Tverd Tela* **5**, 1170 (1963)
217 M G A Bernard and G Duraffourg, *Phys. Status Solidi* **1**, 669 (1961)
218 D N Nasledov, A A Rogachev, S M Ryvkin and B V Tsarenkov, *Sov. Phys.-Solid State* **4**, 782 (1962)
219 As reported by R N Hall, *IEEE Trans. Electron Devices* **ED-23**, 700 (1976)
220 P W Dumke, *Phys. Rev.* **127**, 1559 (1962)
221 At the *Solid State Devices Research Conference, July 1962* R J Keyes and T M Quist reported that at 77 K, a quantum efficiency close to 100% could be achieved for electroluminescence in GaAs; see also *Proc. IRE* **50**, 1822 (1962)
222 J I Pankove and M J Massoulié, *J. Electrochem. Soc.* **109**, 67C (1962); see also J I Pankove, *Phys. Rev. Lett.* **9**, 283 (1962); J I Pankove and J E Berkeyheiser, *Proc. IRE* **50**, 1976 (1962)
223 R N Hall, G E Fenner, J O Kingsley, T J Soltys and R O Carlson, *Phys. Rev. Lett.* **9**, 366 (1962)
224 M I Nathan, W P Dumke, G Burns, F H Dill jr and G Lasher, *Appl. Phys. Lett.* **1**, 62 (1962)
225 R J Keyes and T M Quist, *Proc. IRE* **50**, 1822 (1962); T M Quist, R H Rediker, R J Keyes, W E Krag, B Lax, A L McWhorter and H J Zeiger, *Appl. Phys. Lett.* **1**, 91 (1962)
226 N Holonyak jr and S F Bevacqua, *Appl. Phys. Lett.* **1**, 82 (1962)
227 N Holonyak jr, S F Bevacqua, C V Bielan and S J Lubowski, *Appl. Phys. Lett.* **3**, 47 (1963)
228 A L McWhorter, H J Zeiger and B Lax, *J. Appl. Phys.* **34**, 235 (1963), paper received 23 October 1962
229 Other treatments followed immediately; see for example A Yariv and R C C Leite, *Appl. Phys. Lett.* **2**, 55 (1963); A L McWhorter, *Solid State Electron.* **6**, 417 (1963); G J Lasher, *IBM J.* **7**, 58 (1963); T M Quist, R J Keyes, W E Krag, B Lax, A L McWhorter, R H Rediker and H J Zeiger, *Quantum Electronics* edited by P Grivet and N Bloembergen (Dunod: Paris, 1964) p1833
230 W L Bond, B G Cohen, R C C Leite and A Yariv, *Appl. Phys. Lett.* **2**, 57 (1963)

231 This was achieved by J C Dyment and L A D'Asaro, *Appl. Phys. Lett.* **11**, 292 (1967) who, using an adequate heat sink, obtained CW operation at temperatures up to 205 K

232 H Kroemer, *Proc. IEEE* **51**, 1782 (1963)

233 Zh I Alverov, and R F Kazarinov quoted in Zh I Alverov *et al*, *Sov. Phys.-Solid State* **9**, 208 (1967)

234 I Hayashi, M B Panish and P W Foy, *IEEE J. Quantum Electron.* **QE-5**, 211 (1969); M B Panish, I Hayashi and S Sumski, ibid. **QE-5**, 210 (1969); I Hayashi and M B Panish, *J. Appl. Phys.* **41**, 150 (1970); see also H C Casey jr and M B Panish, *Heterostructure Lasers* (Academic: New York, 1978) part 1 ch 1 where a good historical introduction to semiconductor lasers is provided

235 H Kressel and H Nelson, *RCA Rev.* **30**, 106 (1969)

236 Zh I Alverov V M Andreev, E L Portnoi and M K Trukan, *Sov. Phys.-Semicond.* **3**, 1107 (1970)

237 P L Kapitza and P A M Dirac, *Proc. Camb. Phys. Soc.* **29**, 297 (1933)

238 R H Pantell, G Soncini and H E Puthoff, *IEEE J. Quantum Electron.* **4**, 905 (1968)

239 J M J Madey, *J. Appl. Phys.* **42**, 1906 (1971)

240 L R Elias, W M Fairbank, J M J Madey, H A Schwettman and T I Smith, *Phys. Rev. Lett.* **36**, 717 (1976); *Phys. Today* February 1976 p17

241 D A G Deacon, L R Elias, J M J Madey, G J Ramian, H A Schwettman and T I Smith, *Phys. Rev. Lett.* **38**, 892 (1977); *Sci. Am.* June 1977 p 63

242 R V Palmer, *J. Appl. Phys.* **43**, 3014 (1972)

243 V P Sukhatme and P A Wolff, *J. Appl. Phys.* **44**, 2331 (1973)

244 Some general reviews can be found in *Novel Sources of Coherent Radiation* edited by S F Jacobs, M O Scully and M Sargent III (Addison-Wesley: London, 1978). The problem also has a strict connection with the production of synchrotron radiation (see for example A A Sokolov and J M Ternov, *Synchrotron Radiation* (Pergamon: New York, 1968)) which was first calculated by D Ivanenko and I Pomeranchuk, *Phys. Rev.* **65**, 343 (1944) and T Schwinger, *Phys. Rev.* **70**, 798 (1946); **75**, 1912 (1949), and experimentally verified by F R Elder, R V Langmuir and H C Pollock, *Phys. Rev.* **74**, 52 (1948). Synchrotron radiation is nowadays a very important and powerful source of short-wavelength radiation and is receiving great attention (see for example *Phys. Today* May 1981; Y Farge, *Appl. Opt.* **19**, 4021 (1980))

7

The statistical properties of light

7.1 Introduction

The most fundamental property of laser radiation is its coherence. The very high spectral brightness, the monochromaticity and the directionality of a laser beam are all properties connected with the coherence of its emission. Spectral narrowing and directionality are manifestations of what is nowadays called *first-order coherence*. However, the physical possibility of obtaining both a very high brightness and a much narrower linewidth than the Doppler-broadened line is due to the circumstance that the light emission in a laser takes place in conditions of thermodynamic non-equilibrium. It was consideration of the properties of radiation emitted in conditions of thermodynamic non-equilibrium or, what amounts to the same thing, in conditions in which stimulated emission dominates, that led R J Glauber (1925–) to extend the concept of classical coherence, thereby disclosing the particular statistical properties of radiation emitted by an ideal laser. These peculiar, statistical properties are perhaps the most important general feature of lasers; therefore they deserve special mention in this book.

7.2 The introduction of the concept of the photon

Our story will not concern itself with the old dispute among scientists about the nature of light – the wave theory which was proposed in the 17th century by R Hooke (1635–1703) and C Huygens (1629–1695) versus the corpuscular theory put forward by Isaac Newton (1642–1727). The steps leading to the first quantum theory introduced by Planck and to the concept of quantisation of energy can be found in the excellent book by E Whittaker[1]. The formation of the quantum concept is also well illustrated and discussed by M Jammer[2].

The black-body radiation law was discovered by Max Karl Ernst Ludwig Planck (1858–1947) in 1900[3], just a few months after the publication of the

Rayleigh formula[4]. Within a few weeks Planck observed that, in order to derive his formula, it was necessary to introduce the notion of a *quantum of energy*, which represents the smallest amount of energy that an oscillator can either emit or absorb[5].

According to Planck's theory, an oscillator of frequency ν can emit or absorb energy only in multiples of $h\nu$. Planck regarded the quantum property as belonging essentially to the interaction between radiation and matter: free radiation he supposed to consist of electromagnetic waves, in accordance with Maxwell's theory.

The next important advance was made by Einstein[6] in 1905, who clearly recognised Planck's discovery, which had until then attracted little attention. The paper[6] entitled *On a heuristic point of view concerning the creation and conversion of light* was published in the same volume (17) of *Annalen der Physik* in which the theory of Brownian motion and the special theory of relativity were published[7]: which probably makes this volume one of the most remarkable in the whole scientific literature, as Max Born noted!

This paper is usually referred to nowadays as the Einstein paper on the photoelectric effect[8], but in fact carries a much broader relevance. In it Einstein deduced, from statistical thermodynamics, that the entropy of radiation described by Wien's distribution law has the same form as the entropy of a gas of independent particles. Einstein used this result to argue for the heuristic viewpoint that light consists of quanta, each possessing an amount of energy $\epsilon = h\nu$, and then applied this conclusion to explain the photoelectric effect (indicated as the emission of cathode rays through the illumination of solid bodies). He observes:

> The wave theory of light, which operates with continuous spatial functions, has worked well in the representation of purely optical phenomena and will probably never be replaced by another theory. It should be kept in mind, however, that the optical observations refer to time averages rather than instantaneous values. In spite of the complete experimental confirmation of the theory as applied to diffraction, reflection, refraction, dispersion, etc, it is still conceivable that the theory of light which operates with continuous spatial functions may lead to contradictions with experience when it is applied to the phenomena of emission and transformation of light.
>
> It seems to me that the observations associated with black-body radiation fluorescence, the production of cathode rays by ultraviolet light, and other related phenomena connected with the emission or transformation of light are more readily understood if one assumes that the energy of light is discontinuously distributed in space. In accordance with

the assumption to be considered here, the energy of a light ray spreading out from a point source is not continuously distributed over an increasing space, but consists of a finite number of energy quanta which are localised at points in space, which move without dividing, and which can only be produced and absorbed as complete units.

Einstein's view was, in fact, in contrast with that of Planck, that radiation was quantised not only in the emission process but that it remained quantised also in the propagation process[9].

Einstein uses the word *quantum* of light or of energy. The name *photon* was introduced much later in 1926 by G N Lewis[10] (1875–1946), who thought it inappropriate to speak of a quantum of light:

...if we are to assume that it spends only a minute fraction of its existence as a carrier of radiant energy, while the rest of the time it remains an important structural element within the atom... I therefore take the liberty of proposing for this hypothetical new atom which is not light but plays an essential part in every process of radiation, the name photon.

The explanation of the photoelectric effect in terms of photons took, however, a considerable time to be fully accepted. At the time a resonance theory due to Lenard[11] was commonly accepted.

Einstein's theory was verified in 1912 by O W Richardson (1879–1959) and K T Compton (1881–1954)[12] and by A L Hughes[13] and with great care by R A Millikan (1868–1958)[14] who started by completely disbelieving it and instead gave the best experimental confirmation of it, through 10 years work, being awarded the Nobel Prize in 1923 partly for this.

Surprisingly, no mention or connection with the 'softening' of x-rays scattered by a substance of low atomic weight was made in the discussion. This effect, discovered by C A Sadler and P Mesham in 1912[15], was eventually experimentally confirmed and explained by A H Compton (1892–1962) in 1922[16], and it is nowadays one of the better known examples of the corpuscular behaviour of light[17].

This phenomenon soon became known as the Compton effect, and for this discovery Compton received the Nobel Prize in 1927 sharing it with C T R Wilson (1869–1959) the inventor of the famous cloud chamber.

To get an insight into the difficulties encountered by Einstein's theory of the photoelectric effect it may help to recall the following two examples also reported by M J Klein[8].

Still in 1913 in a letter which proposed Einstein for membership in the Prussian Academy and for a Research Professorship and where Einstein's work was fully appreciated, Planck wrote:

> That he may sometimes have missed the target in his speculations, as for example in his hypothesis of light quanta, cannot really be held against him

(cf C Seelig *Albert Einstein: A Documentary Biography* (London, 1956) pp143–5 reported by M J Klein in *The Natural Philosopher* **2**, 59 (1963).) Some years later in 1916, describing his experimental confirmation of the Einstein equation for the photoelectric effect, Millikan wrote of the same hypothesis (R A Millikan, *The Electron* (Chicago, Ill, 1917) p 238).

> I shall not attempt to present the basis for such an assumption, for, as a matter of fact, it had almost none at the time.

Eventually Albert Einstein was awarded the Nobel Prize in 1921 just for his theory of the photoelectric effect!

Finally, it may be observed that it is commonly believed that Einstein developed his hypothesis of light quanta as an extension of Planck's theory of black-body radiation. That belief is not supported by a careful reading of the work of both physicists, as M Klein has shown[8]. Rather, Einstein postulated the existence of quanta of light on pure thermodynamic grounds (he had previously carefully studied thermodynamics and statistical mechanics writing three works on these subjects) without using Planck's distribution law or his discrete, quantised oscillator energies in his own arguments. He also wrote the magnitude of his light quanta as $(R/N_0)\,\beta\nu$, R being the gas constant, N_0 Avogadro's number and β the exponential coefficient appearing in Wien's radiation formula, and not as $h\nu$.

7.3 Fluctuations of radiant energy

In 1909, four years after his 'photoelectric paper', Einstein published a paper[18] in which he showed that Planck's radiation law itself implies that the radiation field exhibits not only wave features but also corpuscular features. This result, which was the first clear indication of the so-called *wave-particle duality*, has notable importance in our history.

In this paper Einstein calculated the fluctuations of radiant energy in a partial volume V of an isothermal enclosure at temperature T. The subject of radiation fluctuation played a key role in the development of quantum statistics which followed in the period up to 1930. Later, much interest was centred on the practical problem of attaining the maximum detectivity in spectrometric investigations.

Like other problems of fluctuations connected with departure from equilibrium, radiation fluctuations in a cavity can be treated with the methods of statistical mechanics.

By denoting by E the instantaneous energy in the cavity within the frequency interval ν to $\nu + \mathrm{d}\nu$, with an average value

$$\bar{E} = \int E \exp(-E/kT)\, \mathrm{d}p\, \mathrm{d}q \Big/ \int \exp(-E/kT)\, \mathrm{d}p\, \mathrm{d}q, \tag{7.1}$$

and with $\epsilon = E - \bar{E}$ the energy fluctuation, Einstein was easily able to show that the mean-square fluctuation $\overline{\epsilon^2}$, which is defined as

$$\overline{\epsilon^2} = \overline{E^2} - \bar{E}^2, \tag{7.2}$$

where

$$\overline{E^2} = \int E^2 \exp(-E/kT)\, \mathrm{d}p\, \mathrm{d}q \Big/ \int \exp(-E/kT)\, \mathrm{d}p\, \mathrm{d}q, \tag{7.3}$$

is given by

$$\overline{\epsilon^2} = kT^2\, \mathrm{d}\bar{E}/\mathrm{d}T. \tag{7.4}$$

The mean energy $\bar{E}$ can also be obtained by Planck's law; substituting in

$$\bar{E} = V\rho_\nu\, \mathrm{d}\nu, \tag{7.5}$$

the expression for ρ_ν given by Planck, Einstein found

$$\begin{aligned} \overline{\epsilon^2} &= (8\pi h^2\nu^4\, \mathrm{d}\nu/c^3)\{1/[\exp(h\nu/kT) - 1] + 1/[\exp(h\nu/kT) - 1]^2\} \\ &= h\nu E + c^3E^2/8\pi\nu^2\, \mathrm{d}\nu. \end{aligned} \tag{7.6}$$

Einstein observed that if, instead of Planck's law of radiation, one had taken Wien's law, one should have obtained

$$\overline{\epsilon_W^2} = h\nu E; \tag{7.7}$$

while if one had taken Rayleigh's law one should have obtained

$$\overline{\epsilon_R^2} = c^3E^2/8\pi\nu^2\, \mathrm{d}\nu. \tag{7.8}$$

Therefore, the mean-square value of the fluctuations according to Planck's law is the sum of the mean squares of the fluctuations according to Wien's law and Rayleigh's law. The result, looked at in the light of the principle that fluctuations due to independent causes are additive, suggests that the causes operative in the case of high frequencies (for which Wien's law holds) are independent of those operative in the case of low frequencies (for which Rayleigh's law is valid). Now, Rayleigh's law is based on the wave theory of light, and, in fact, Lorentz[19] showed that the value $c^3E^2/8\pi\nu^2\, \mathrm{d}\nu$ is a consequence of the interference of wave trains, which, according to the classical picture, are crossing the cavity in every direction; whereas the value $h\nu E$ for the mean-square fluctuations is what would be obtained if one were to take the formula for the fluctuations of the number of molecules in a unit volume of an ideal gas and suppose that each molecule has energy $h\nu$ – that is, the expression which would be obtained by a corpuscular quantum theory.

Moreover, the ratio of the particle term to the wave term in the complete expression for the fluctuation is

$$\exp(h\nu/kT) - 1.$$

When $h\nu/kT$ is small, i.e. at low frequencies and high temperatures, the wave term is predominant; and when $h\nu/kT$ is large, i.e. when the energy density is small, the particle term predominates.

The formula therefore suggests that light cannot be completely represented either by waves or by particles, although for some phenomena the wave representation is practically sufficient, and for other phenomena the particle representation is good.

Einstein was greatly attracted by the problem of light, and later, in 1915-16, published the paper in which he introduced the concept of stimulated emission and which once again was based on the corpuscular concept of light[20, 21]. Meanwhile, the photon hypothesis was carefully examined by several people. P Ehrenfest[22], A Joffé[23], L Natanson[24] and G Krutkow[25] showed that if we assume that each of the light quanta of frequency ν has an energy $h\nu$ and that they are completely independent of each other, Wien's law of radiation is obtained instead of Planck's law.

In order to obtain Planck's formula it is necessary to assume that elementary photons of energy $h\nu$ form aggregates of energies $2h\nu$ and $3h\nu$, respectively, and that the total energy of radiation is distributed, on average, in a regular manner between them. L de Broglie (1892-), later (in 1922)[26, 27] made a similar remark concerning the energy fluctuations. He observed that Planck's formula

$$E = (8\pi h\nu^3/c^3)\{\mathrm{d}\nu/[\exp(h\nu/kT) - 1]\} \tag{7.9}$$

may be written

$$\begin{aligned} E &= (8\pi h\nu^3/c^3)[\exp(-h\nu/kT) + \exp(-2h\nu/kT) + \exp(-3h\nu/kT) + \ldots]\,\mathrm{d}\nu \\ &= E_1 + E_2 + E_3 + \ldots + E_s + \ldots \end{aligned} \tag{7.10}$$

where

$$E_s = (8\pi h\nu^3/c^3)\exp(-sh\nu/kT)\,\mathrm{d}\nu. \tag{7.11}$$

Now, Einstein's formula[6] can be written

$$\begin{aligned} \overline{\epsilon^2} &= (8\pi h^2\nu^4\,\mathrm{d}\nu/c^3)[\exp(-h\nu/kT) + 2\exp(-2h\nu/kT) + 3\exp(-3h\nu/kT) + \ldots] \\ &= \sum_{s=1}^{\infty} sh\nu E_s. \end{aligned} \tag{7.12}$$

This resembles the first term $h\nu E$ in formula (7.6), but it is now summed for all values of s. So it is precisely the result we should expect if the energy E_s was made up of light quanta, each of energy $sh\nu$. Thus de Broglie suggested that the term E_1 should be regarded as corresponding to energy existing in the form of quanta of amount $h\nu$; that the second term E_2 should be regarded as corresponding to energy existing in the form of quanta of amount $2h\nu$; and so on.

So, the Einstein formula for the fluctuations may be obtained on the basis of a purely corpuscular theory of light, provided the total energy of the radiation is suitably allocated among corpuscles of different energies $h\nu$, $2h\nu$, $3h\nu$, ...

The following year W Bothe[28] gave the calculation of the number of quanta $h\nu$ in black-body radiation which are associated as 'photo-molecules' in pairs, $2h\nu$, trios, $3h\nu$, etc which has already been discussed in Chapter 3[29]. These associations are responsible for the bunching properties of radiation we shall discuss below.

7.4 Bose and the statistics of radiation

A further step forward was made by Satyendra Nath Bose[30] in 1924 at Dacca University, India in a short paper wherein a new proof of Planck's formula was given.

Bose was born in 1894 and died on 4 February 1974. In 1916 he and M N Saha (1893–1956) became two of the first lecturers in the new University College of Science built in 1914 by Vice-Chancellor Sir Asutosh Mookerjee in Calcutta, a college where the professorial posts had to be filled by Indians. The following year, with the creation of a department, lecturers in physics were nominated.

Bose immediately showed his interest in statistical physics. In 1919, together with Saha, he published an anthology of works by Einstein on relativity, one of the first of such collections published in English. In 1921 Bose left Calcutta to become Reader in Physics in the reorganised University of Dacca in East Bengal. There he lectured, read, thought, and spent many sleepless nights thinking about Planck's law. In late 1923 he submitted a paper on the subject of the prestigious English review, *Philosophical Magazine*. Six months later the editors of that magazine informed him that the referee's report on his paper was unfavourable.

He then sent the manuscript to Einstein (on 4 June 1924), who was then Professor of Physics at Berlin. The paper was accompanied by a letter which began[31]:

> Respected Sir,
>
> I have ventured to send you the accompanying article for your perusal and opinion. I am anxious to know what you think of it. You will see that I have tried to deduce the coefficient $8\pi\nu^2/c^3$ in Planck's law independent

> of the classical electrodynamics, only assuming that the ultimate elementary region in the phase space has the content h^3. I do not know sufficient German to translate the paper. If you think the paper worth publication I shall be grateful if you arrange for its publication in *Zeitschrift für Physik*. Though a complete stranger to you, I do not hesitate in making such a request. Because we are all your pupils though profiting only from your teachings through your writings . . .

Einstein translated the paper and sent it in July 1924 to the *Zeitschrift* in Bose's name, where it was published under the title *Plancks Gesetz und Lichtquantenhypothese* (*Planck's law and the hypothesis of light quanta*)[30]. He also added a note stating:

> In my opinion Bose's derivation of the Planck formula signifies an important advance. The method used also yields the quantum theory of the ideal gas, as I will work out in detail elsewhere.

He then sent a postcard to Bose signifying that he considered the work a most important contribution.

Bose's method is based on arguments relative to phase space. He considered the radiation as being composed of photons, which for statistical purposes can be treated as particles of a gas, with the important difference that photons are indistinguishable one from the other; so that, instead of considering the disposition of photons as being individually distinguishable in an ensemble of states, he paid attention to the number of states which contain a given number of photons. He assumed that the total energy E of the photons was known and that they were enclosed in a cavity of unit volume.

The frequency distribution of the radiation at an absolute temperature T was then deduced by finding the distribution in phase space that maximises the entropy of the system.

To this end Bose observed that a photon $h\nu$ can be specified by its coordinates (x, y, z) and by the three components of its momentum (p_x, p_y, p_z). Because the total momentum is $h\nu/c$, we have

$$p_x^2 + p_y^2 + p_z^2 = r^2, \qquad \text{where} \qquad r = h\nu/c.$$

By using a phase space of six dimensions (x, y, z, p_x, p_y, p_z), for the frequency range ν to $\nu + \mathrm{d}\nu$, the corresponding volume is

$$\int \mathrm{d}x\ \mathrm{d}y\ \mathrm{d}z\ \mathrm{d}p_x\ \mathrm{d}p_y\ \mathrm{d}p_z = 4\pi r^2\ \mathrm{d}rV = 4\pi V(h^3\nu^2/c^3)\ \mathrm{d}\nu. \qquad (7.13)$$

Bose now assumed that this space is divided into cells, each of volume h^3, so that in it there are $(4\pi\nu^2\ \mathrm{d}\nu/c^3)\ V$ cells. To take account of polarisation one must then multiply this number by two.

Now, let N_s be the number of quanta belonging to the frequency domain $d\nu^s$ and then let us consider the number of ways in which these quanta can be distributed among the cells belonging to $d\nu^s$. Let p_0^s be the number of vacant cells; p_1^s the number of those which contain one quantum; p_2^s the number of cells containing two quanta, etc; then the number of possible ways of choosing an ensemble of p_0^s, p_1^s cells, etc from the total of $8\pi\nu^2\, d\nu/c^3$ cells is

$$A^s!/p_0^s!\,p_1^s!\,p_2^s!\ldots,$$

where

$$A^s = (8\pi\nu^2/c^3)\, d\nu^s,$$

and we have

$$N^s = \sum_r r p_r^s.$$

As a fundamental hypothesis of his statistics, Bose now assumed that if one considers a particular quantum state, all the values for the number of particles in that state are equivalent, so that the probability of each distribution specified by p_r^s is measured by the number of different ways in which it can be realised. Therefore the probability of a state specified by p_r^s (taking into account all the intervals of frequency) is

$$W = \prod_s \cdot A^s!/p_0^s!\,p_1^s!\,p_2^s!\ldots.$$

Because the p_r^s are large, one may use Stirling's approximation: $\log n! = n \log n - n$. So

$$\log W = \sum_s A^s \log A^s - \sum_s \sum_r p_r^s \log p_r^s,$$

where

$$A^s = \sum_r p_r^s.$$

This expression should be a maximum satisfying the auxiliary conditions

$$E = \sum_s N^s h\nu^s \qquad N^s = \sum_r r p_r^s.$$

Carrying out the variation gives the conditions

$$\sum_s \sum_r \delta p_r^s (1 + \log p_r^s) = 0 \qquad \sum_s \delta N^s h\nu^s = 0$$

$$\sum_r \delta p_r^s = 0 \qquad \delta N^s = \sum_r r\delta p_r^s.$$

It follows that

$$\sum_s \sum_r \delta p_r^s [(1 + \log p_r^s + \lambda^s) + (rh\nu_s/\beta)] = 0,$$

where β and λ^s are constant, so

$$p_r^s = B_r^s \exp(-rh\nu^s/\beta),$$

where B^s are constant.

Therefore

$$A_s = \sum_r p_r^s = \sum_r B^s \exp(-rh\nu^s/\beta) = B^s[1 - \exp(-h\nu^s/\beta)]^{-1},$$

or

$$B^s = A^s[1 - \exp(-h\nu^s/\beta)],$$

while

$$N_s = \sum_r rp_r^s = A^s \sum_r r \exp(-h\nu^s/\beta)[1 - \exp(-h\nu^s/\beta)]$$

$$= A^s \exp(-h\nu^s/\beta)/[1 - \exp(-h\nu^s/\beta)].$$

Therefore

$$E = \sum_s N_s h\nu_s = \sum_s [8\pi h(\nu^s)^3/c^3] \exp(-h\nu^s/\beta)[1 - \exp(-h\nu^s/\beta)]^{-1} \, d\nu^s V. \tag{7.14}$$

Using the preceding results, one also finds that the entropy is

$$s = k \log W = k\left((E/\beta) - \sum_s A^s \log[1 - \exp(-h\nu^s/\beta)]\right). \tag{7.15}$$

Because

$$\partial s/\partial E = 1/T,$$

where T is the absolute temperature, one has $\beta = kT$. Therefore

$$E = \sum_s [8\pi h(\nu^s)^3/c^3][\exp(h\nu^s/kT) - 1]^{-1} \, d\nu^s, \tag{7.16}$$

which is equivalent to Planck's formula.

Bose's work showed that to obtain Planck's law one has to suppose that photons obey a given kind of statistics[32].

Bose's discovery was immediately extended by Einstein to the study of a monatomic ideal gas[33]. His analysis brought the following conclusions: the average number of particles of mass m in a unit volume with energies between ϵ and $\epsilon + d\epsilon$ is

$$d\epsilon\bar{n} = (2\pi/h^3)(2m)^{3/2}\epsilon^{1/2}\, d\epsilon/[\exp(\epsilon/kT + \mu) - 1] \qquad (7.17)$$

where μ is a constant, the total number of particles per unit volume is

$$N = (2\pi/h^3)(2m)^{3/2}\int_0^\infty \epsilon^{1/2}\, d\epsilon/[\exp(\epsilon/kT + \mu) - 1]$$

and the total energy per unit volume is

$$E = (2\pi/h^3)(2m)^{3/2}\int_0^\infty \epsilon^{3/2}\, d\epsilon/[\exp(\epsilon/kT + \mu) - 1]. \qquad (7.18)$$

These are the fundamental formulae of what is generally called Bose-Einstein statistics. Since equation (7.17) differed from the Boltzmann factor $\exp(-\epsilon/kT)$ of ordinary statistical mechanics, all the thermodynamic properties of the gas were correspondingly more complicated. Einstein was able to show, however, that when the temperature was high and the density was low his equations returned to those for the classical gas.

In 1924 Bose received support for a study period of two years in Europe and arrived in Paris in September. He recounts that on the strength of the postcard Einstein had sent him, the German Consulate in Calcutta issued his visa without requiring payment of the customary fee!

On arrival in Paris he met P Langevin (1872-1946), who suggested to him the possibility of working in Mme Curie's laboratory. Bose recalled this meeting with amusement in an interview with W A Blanpied[31]. Mme Curie spoke in English all the time and did not let him say a single word.

She spoke to him about another Indian student who worked with her and had encountered serious difficulties because he did not speak good French. Then she suggested that Bose should concentrate on the language for six months and then come back to her. Bose did not even get the chance to tell her that he had studied French for 15 years!

After this discouraging contact Bose met the de Broglie brothers and was with Maurice for some time. However, he was still very anxious to go to Einstein. On 26 October 1924 he had written Einstein a letter which began:

Dear Master

My heartfelt gratitude for taking the trouble of translating the paper yourself and publishing it. I just saw it in print before I left India. I have sent you about the middle of June a second paper entitled *Thermal Equilibrium in the Radiation Field in the Presence of Matter.*

I am rather anxious to know your opinion about it as I think it to be rather important. I don't know whether it will be possible also to have this paper published in *Zeitschrift für Physik.*

I have been granted study leave by my university for two years. I have arrived just a week ago in Paris. I don't know whether it will be possible for me to work under you in Germany. I shall be glad however, if you grant me the permission to work under you, for it will mean for me the realisation of a long cherished hope.

I shall wait for your decision as well as your opinion of my second paper here in Paris. . . .

Einstein had already translated this second paper and sent it to the *Zeitschrift* which published it[34]. This time, however, Einstein added a remark stating that he could not agree with his author's conclusions, and went on to give his reasons[35]. Apparently he communicated these objections privately because in a letter from Paris dated 27 January 1925, Bose thanks Einstein for his communication of 3 November and states he is about to send off to him another paper which he hopes will satisfy his objections. This work, however, was never published nor can any other mention of it be found in Bose's correspondence. There is no copy of Einstein's letter available and so his plans for Bose's visit in Berlin are unknown, however, Bose had not given up the idea of coming, for in the letter dated 27 January, he closes by again stating that he hopes to be able to work with Einstein.

Nowadays, it may be said that the opportunity of a collaboration between Bose and Einstein had already evaporated by January 1925.

In July 1924, about the time Bose was finally settling the study leave question with the Dacca University authorities, Einstein was reading a paper[33] before the Prussian Academy where he applied Bose's statistical method to an ideal gas with particles of non-zero rest mass and suggested some of its non-chemical thermodynamic properties.

The similarity in the statistical behaviour between photons of Bose and gas particles, found in the paper[33] was further investigated by Einstein. This non-chemical gas continued to fascinate him for several months. In September 1924 (while Bose was at sea), Einstein wrote to Ehrenfest stating that at low temperatures the molecules of the gas would condense into the zero-energy state even in the absence of attractive forces between them[36].

At the beginning of December (while Bose was pondering Einstein's objections to his second work) Einstein wrote to Ehrenfest 'The matter of the quantum gas is getting very interesting'[36,37].

Finally, in January 1925, Einstein published a second paper[38] where he observed that in expression (7.6) both terms appear if one assumes that particles satisfy Bose statistics, and made clear that the particles were not treated as independent in the Bose-Einstein counting procedure. He derived the analogue of equation (7.6) as

$$(\Delta s)^2 = n_s + n_s^2/z_s$$

where Δs was the fluctuation about the average value n_s of the number of molecules in a specified energy interval, and z_s is the number of cells. He also predicted the now well known Bose-Einstein condensation and tried to rationalise Bose's method, after which he turned his attention to other matters[39].

When Bose arrived in Berlin, and on 8 October 1925 wrote to Einstein asking for an appointment, Einstein had ceased to be interested in the argument and was in Leyden. He came back only after several weeks. When finally the two men met, the encounter was very disappointing. As a result Bose obtained a letter allowing him to enjoy some privileges common to students in Berlin, including permission to take books from the university library!

Later, in the summer of 1926 Bose returned to Dacca and, one year later, was appointed Physics Professor. He held this position until 1945 when he went back to Calcutta University, again as Physics Professor.

During all those years in Dacca he worked and dedicated himself to Indian and Western philosophy and literature. He also interested himself in politics and was an ardent supporter of the campaign for Bengali independence.

He returned to France in 1951 and was often abroad, but he never wanted to go back to Germany. He occupied himself with topics connected with Einstein. Of the six papers he published between 1953 and the time of his retirement, five treat the unified field theory of Einstein. The last of Einstein's letters to him is dated 22 October 1953, containing thanks for a preprint received and for the enclosed friendly letter[31]. It can be stated, however, that after the paper on statistics Bose did not contribute any more to the forefront of physics. He had arrived in Europe in 1924, met Langevin and de Broglie in a period when the new physics was boiling: so why did he lose the opportunity of participating? Partly, perhaps, because of his placid temperament: Einstein understood everything on the quantum gas in seven months. Partly also because he was intimidated by most Europeans. The heavy contribution that Einstein had given to the determination of Bose statistics was soon recognised, and the statistics were named Bose-Einstein. Dirac, who first coined the word 'boson', seems to have been among the few European physicists who gave Bose full credit for his achievement, as Bose himself noted in an interview given three years before his death[31].

7.5 Further developments in the theory of fluctuations of radiation fields

In the years following, up to about 1930, the subject of radiation fluctuations was extensively considered by A Smekal[40], W Bothe[41] and R Furth[42, 43]. Furth[43] performed an important generalisation, showing that equation (7.6) can be obtained directly by the application of Bose-Einstein statistics to photons, irrespective of their spectral distribution. In these papers two expressions for the statistics appear. The probability that a state with energy E is occupied as

$$W(E) = 1/[\exp(E/kT) - 1], \tag{7.19}$$

and the probability $p(n)$ of having n photons of a distribution with mean value $\bar{n}$ as

$$p(n) = 1/[(1 + \bar{n})(1 + 1/\bar{n})^n]. \tag{7.20}$$

All these treatments were related to closed systems rather than to propagating light beams.

In the 1940s, during the war, the development and application of radiation detectors focused attention on the importance of investigating the fluctuations of electromagnetic fields propagating far from their sources.

The first studies by Burgess[44, 45] were focused on fluctuations of radio waves and noise induced in radio antennas.

Other works[46–49] on a large variety of radiation detectors showed that the fluctuations of the radiation field imposed an ultimate limit on the measurements in every case.

In particular, it was shown by Burgess[44] and later by others[50–52] that the spectral density of the radiation fluctuations obeyed the quantum form of the well known Nyquist formula for Johnson's noise[53].

Lewis[47], Fellgett[49] and Jones[54] also obtained expressions by a thermodynamic argument for the mean-squared fluctuation $\overline{(\Delta n)^2}$ of the number of photon counts n recorded by an illuminated photocell. Although some of these expressions were later shown to be inapplicable, after some discussion[55–57] the fact was recognised that the fluctuations of n depart from the classical counting statistics and that $\overline{(\Delta n)^2}$ can be expressed as the sum of two contributions in close analogy with equation (7.6).

In 1948, in a review article, MacDonald[58] considered it '... clear *a priori* that no information about the energy distribution of the incident photons can be derived ...' from counting measurements with a photocell. This is nowadays known to be incorrect.

Investigations into coherence as a manifestation of correlations between interfering fields began somewhat earlier, in the 19th century[59]; the historical path these investigations took has been followed in papers by L Mandel and

E Wolf[60], and M Born and E Wolf[61]. The main outcome of this research before the 1950s was twofold: the introduction of a precise measure of correlation between the field variables at two space-time points; and the formulation of dynamical laws which the correlations obey.

The *classical coherence theory* which is now well established is set up by introducing an analytical representation of the field variables (e.g. the electric field of the wave) through the complex function $V(t)$, known as the *analytic signal*[62]. Any correlation experiment due to interference could, before the 1950s, be treated by introducing the correlation function

$$\Gamma(r_1, r_2, \tau) = \langle V^*(r_1, t)\, V(r_2, t+\tau)\rangle = \Gamma_{12}(\tau)$$

$$= \lim_{T\to\infty} (1/2T) \int_{-T}^{+T} V^*(r_1, t)\, V_2(r_2, t+\tau)\, \mathrm{d}t. \tag{7.21}$$

The ensemble correlation function Γ was first introduced by Zernike (1888–1966) in 1938[63] (although not of course in terms of the analytic signal), together with its normalised quantity

$$\gamma_{12}(\tau) = [\Gamma_{12}(\tau)]\,[\Gamma_{11}(0)]^{-1/2}\,[\Gamma_{22}(0)]^{-1/2} \tag{7.22}$$

called by him the *complex degree of coherence*. γ_{12} is 1 for a fully coherent source and 0 for a fully incoherent source. The time τ at which $\gamma(\tau)$ reaches the value $1/e$ is called the *coherence time*.

The use of these functions was further developed by Hopkins[64]. Several important properties and extensions of correlation functions were derived by Wolf[65] and, independently, by Blanc-Lapierre and Dumontet[66] in 1955. Some of these are:

(1) The value of γ_{12} for a plane illuminated by a circular source of diameter a, at distance R[64], is

$$\gamma_{12} = [2J_1(z)/z] \exp[ik(\rho^2/2R)], \tag{7.23}$$

where $J_1(z)$ is the Bessel function of the first kind and first order and

$$z = (2\pi/\lambda) \sin(\rho^2/R); \tag{7.24}$$

(2) The demonstration of the fact that the coherence factor is essentially the normalised integral over the source of the Fourier (frequency) transform of the intensity function of the source[65].

The clarification of the concepts of coherence which resulted from these discussions enabled Forrester to develop an old idea which he had had several years before: the production of beats from light waves[67], or *photoelectric mixing* as he called it. Photoelectric mixing is a direct analogue of the familiar mixing of

AC electrical signals in non-linear circuit elements. In the Forrester *et al* experiment a ^{198}Hg lamp was placed in a magnetic field of 3300 G, and the Zeeman-split 5461 Å light illuminated the photocathode of a microwave phototube. The two strong σ components of the light (for which the Zeeman splitting was about 10^{10} Hz) beat with each other in the square-law photoelectric emission process, and the photocurrent was found to have an AC component at the Zeeman-difference frequency.

The greatest concern in obtaining beats was to get the two waves coherent. Forrester argued that the different spectral components of a homogeneously broadened line should beat between themselves and derived a very simple relation which showed that to obtain a well defined beat pattern it was necessary that the coherence time of the light should be long compared with the period of the beat. Forrester's experiment was very difficult at the time, due to the low intensity and large bandwidth of the light source.

The topic found no immediate application; however, some years later, in 1959, Alkemade[68] pointed out that a single optical spectral line falling on a phototube would produce a photocurrent whose low-frequency noise spectrum would provide information on the shape, and particularly the width, of the optical line. The same conclusion was reached by Forrester in 1961[69] in a paper entitled *Photoelectric Mixing as a Spectroscopy Tool*. He considered a spread spectral line as a series of 'slices', and treated beats between all possible pairs of these slices as a simple extension of the two-component case. With this approach Forrester deduced the photocurrent spectra which would be produced by several different types of optical spectra.

He also discussed the two different approaches that can be used, namely producing beats between two different sources and between different spectral parts of the same line. He called these *superheterodyne* and *low-level detection* respectively[70].

These two approaches were subsequently used to study the spectral output of lasers.[71] They form the basis of the now well known field of *light beating spectroscopy*[72].

7.6 The Hanbury Brown and Twiss experiment

Before the early 1950s, interest in the field of radiation fluctuations was largely based on the unavoidable limitations which they imposed on the accuracy of radiation measurements. However, in the years 1952–6, Hanbury Brown *et al*, working in the field of radio astronomy, developed a new interferometry technique for the determination of angular diameters of radio sources[73–75] in which the measurement of radiation fluctuations played a new and essential role.

Later on they applied this technique to light waves[76–81]. They showed, amongst other things, that the degree of coherence at two points in a radiation field can be inferred from correlation measurements of the fluctuating signals appearing at radio antennas placed at the two points.

Hanbury Brown[82] tells us he had the idea for the intensity interferometer on a night in 1949, while he was designing a radio interferometer to measure the angular dimensions of two important radio sources, Cygnus A and Cassiopeia A. An ordinary radio interferometer would need a base of several thousand km. Would it be possible to construct such an apparatus? He wondered:

> The immediate technical difficulty in modifying a conventional system was to feed a coherent oscillator to the two distant points . . . and I started to wonder if this was really necessary.
>
> Could one perhaps compare the radio waves received at two points by some other means? As an example, I imagined a simple detector which demodulated waves from the source and displayed them as the usual noise which one sees on a cathode-ray oscilloscope. If one could take simultaneous photographs of the noise at two stations, would the two pictures look the same? This question led directly to the idea of the correlation of intensity fluctuations and to the principle of intensity interferometry[83].

After some days of meditation, Hanbury Brown convinced himself that the idea was good and asked the help of Richard Twiss to give him a mathematical basis for his idea. The principle of the interferometer was born and was published later in 1954[74]. In the meantime Hanbury Brown and Twiss, together with R C Jennison, assembled an intensity radio interferometer at Jodrell Bank at a frequency of 125 MHz and used it to measure the diameter of the sun. The next step was, in 1952, the measurement of the radius of Cygnus and Cassiopeia[73]. These measurements showed that the radii of these stars were in fact much larger than expected and could have been measured with a normal interferometer without the large base prepared by Hanbury Brown. This is when something new was observed[84]:

> At the beginning of this programme we had thought that the sole advantage of an intensity interferometer, compared with the radio version of Michelson's interferometer, was that it did not require mutually coherent local oscillators at the separated stations and was therefore peculiarly suitable for extremely long baselines. . . . However, as we watched our interferometer at work, we noticed that when the radio sources were scintillating violently, due to ionospheric irregularities, the measurements of correlation were not significantly affected. Richard Twiss investigated the theory of this surprising effect and confirmed that it was to be expected. We had overlooked one of the principal features of

> an intensity interferometer: the fact that it can be made to work through a turbulent medium.

The team went on to consider the possibility of constructing a light interferometer and Twiss studied its theory. Calculations showed that, to measure a first magnitude star, two telescopes of at least 2.5 m diameter would be required, one of which would need to be mobile.

> It took us six months to realise that although we should certainly need two very large telescopes, they could be extremely crude by astronomical standards

adds Hanbury Brown. In practice, in fact, the telescopes were only needed to collect the light and not to give an excellent quality image as for astronomical telescopes[85]. At this point Hanbury Brown recalls:

> With renewed enthusiasm we returned to establish the detailed theory of an optical interferometer, and immediately ran into a barrage of criticism . . .
>
> Our original theory was clearly correct at radio wavelengths but when it came to light waves there were one or two lingering doubts in our own minds and several firmly entrenched doubts in the minds of others. The trouble of course was due to worrying about photons. As I have already pointed out, radio engineers in those days looked on radio waves as being simply waves and our theory of the radio intensity interferometer was accepted without question. But when we came to deal with physicists, all sorts of queries were raised. One group of objections was concerned with the validity of our semiclassical model of photoelectric emission. We had assumed that the probability of emission of a photoelectron is proportional to the instantaneous square of the electric vector of the incident light wave treated classically. Further, we had assumed that there is no significant delay in the photoelectric process and that in the output current all the components of the envelope of light, at least up to 100 MHz, would be present with their correct phases and amplitudes. At that time there was no sufficiently detailed quantum mechanical treatment of photoelectric emission . . . another stream of objections about photons were both instructive and entertaining. Our whole argument was based on the idea that the fluctuations in the outputs of two photoelectric detectors must be correlated when they are exposed to a plane wave of light. We had shown that this must be so by a semiclassical analysis in which light is treated as a classical wave and in this picture there is no need to worry about photons – the quantisation is introduced by the discrete energy levels in the detector. However, if one must think of light in terms of

photons then, if the two pictures are to give the same result, one must accept that the times of arrival of these photons at the two separated detectors are correlated – they tend to arrive in pairs. Now, to a surprising number of people, this idea seemed not only heretical but patently absurd and they told us so in person, by letter, in publications, and by actually doing experiments which claimed to show that we were wrong. At the most basic level they asked how, if photons are emitted at random in a thermal source, can they appear in pairs at two detectors? At a more sophisticated level the enraged physicist would brandish some sacred text, usually by Heitler, and point out that the number n of quanta in a beam of radiation and its phase Φ are represented by non-commuting operators and that our analysis was invalidated by the uncertainty relation

$$\delta n \cdot \delta \Phi \approx 1. \tag{7.25}$$

We tried as best we could to answer all these objections and to quieten people down. We were certainly interested in seeking the truth but in raising money to build an interferometer it was desirable that our proposals should be widely regarded as sound. These difficulties about photons troubled physicists who had been brought up on particles and had not fully appreciated that the concept of a photon is not a complete picture of light. Thus many people are reluctant to accept the notion that a particular photon cannot be regarded as having identity from emission to absorption. These objections can, in fact, be answered straight out of text-books and we developed some considerable skill in expounding the orthodox paradoxical nature of light, or, if you like, explaining the incomprehensible – an activity closely, and interestingly, analogous to preaching the Athanasian Creed. In answer to the more sophisticated objection that our proposal was inconsistent with the uncertainty relation in equation (7.25) we pointed out that we were proposing to measure only the relative phase $(\Phi_1 - \Phi_2)$ between two beams of radiation; the total energy of two beams $(n_1 + n_2)$ and their relative phase $(\Phi_1 - \Phi_2)$ can be represented by commuting operators and can be represented classically.

Finally, there were the objections based on laboratory experiments which claimed to show that photons are not correlated at two separate detectors. Here we were on sure ground because we had already done our own careful laboratory test of the principle. In 1955, I had borrowed the dark-room which housed the spectro-heliograph at Jodrell Bank and set up our first optical interferometer. An artificial star was formed by focusing the brightest part of a high-pressure mercury arc onto a pinhole. The light from this pinhole was then divided into two beams, by a half-aluminized mirror, to illuminate two photomultipliers mounted so that their photocathodes could be optically superimposed or separated by a variable distance as seen from the pinhole. The whole system [see figure 7.1] simulated the measurement by two detectors on the ground of a star with a surface temperature of about 8000 K. After the usual troubles with the equipment we observed the expected correlation and successfully measured it....

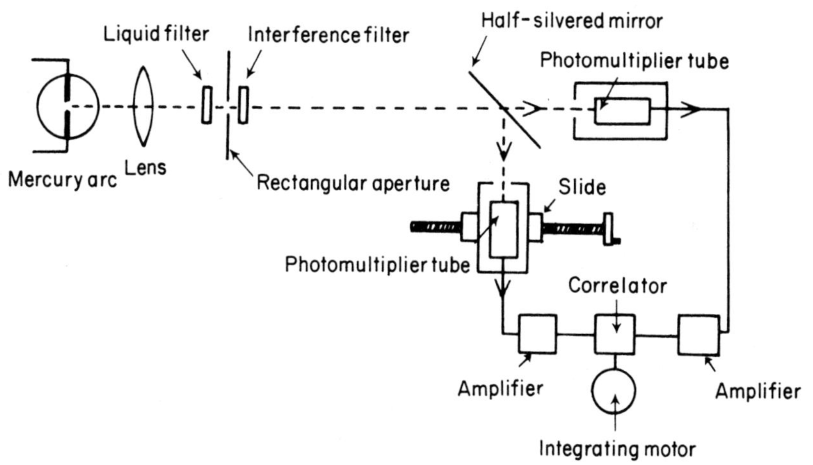

Figure 7.1 Simplified diagram of the apparatus used by Hanbury Brown and Twiss[76]. Reprinted by permission from *Nature* **177**, 27; copyright © 1956 Macmillan Journals Limited.

We were therefore able to face with confidence objections based on two independent experiments claiming to show that there is no correlation between photons in coherent light beams. The first was performed in Budapest by Ádám, Janossy and Varga (1955)[86] and was published at the same time as our own test was being made. In the introduction to their paper they stated that, according to quantum theory, the pulses produced in two separate detectors illuminated by coherent light should be independent of one another. Their aim was 'to investigate the validity of this prediction of quantum theory'. They illuminated two photomultipliers with coherent light from a single source and also with incoherent light from separate sources, and they counted the coincidences of the pulses produced by individual photons in the two phototubes. In an observation lasting ten hours they found no significant correlation between the arrival times of photons and they claimed that this showed that 'in agreement with quantum theory, the photons of two coherent light beams are independent of each other or at least that the biggest part of such photons are independent of each other'. Since the results of this experiment were welcomed by our critics as evidence that an intensity interferometer was fundamentally unsound, we took a closer look at these claims. It was at once obvious, from a quantitative analysis of the parameters of their experiment – light intensity, resolving time, etc – that there was no hope whatever of observing correlation within 10 h or of testing the predictions of quantum theory. In reply we published a brief note[87] drawing attention to the fact that in order to observe a significant correlation (three times RMS noise) Ádám *et al* would have had to observe for 10^{11} years – somewhat longer than the age of the Earth.

The second experimental objection was made in 1956 at the University of Western Ontario by Brannen and Ferguson[88] just after the publication of our own laboratory work. They designed their optical system to resemble as far as possible the one we had used at Jodrell Bank. Two photomultipliers were illuminated by coherent light from a high-pressure mercury arc via a half-silvered mirror and the outputs of the two phototubes were taken, not to a linear multiplier, but to a coincidence counter. They concluded that 'there is no correlation (less than 0.01%) between photons in coherent light rays'. They added that 'if such a correlation did exist it would call for a major revision of some fundamental concepts in quantum mechanics'. Again, we analysed these conclusions and found that the parameters, as before, were hopelessly inadequate to allow the detection of correlation between photons, within a reasonable time. In this case the essential point is that in order to achieve a practical signal-to-noise ratio with a coincidence counter one needs an intense source of light with an extremely narrow bandwidth, and this they did not have. We published a short note[87] showing that it would have taken Brannen and Ferguson 1000 years to observe a significant correlation.

Both these experiments were beyond reproach from an experimental point of view, but since they had been planned without an adequate theoretical foundation they were far too insensitive to be of any significant use. Nevertheless, they did provide our opponents with ammunition.

'Bloody but unbowed' Hanbury Brown and his collaborator assembled the first stellar intensity interferometer by using military relics and were able to measure throughout the winter of 1955–6 the angular diameter of Sirius[77, 81]. The year before they had performed the laboratory test[76] already referred to and another was undertaken in 1957 at the Jodrell Bank Experimental Station of the University of Manchester[79] as Hanbury Brown recalls above, in response to the many objections raised by Ádám *et al*[86] and Brannen and Ferguson[88].

The apparatus used in the first experiment is shown in figure 7.1[76]. Light from a mercury lamp was filtered and split into two beams by a half-silvered mirror and fell on two photocells whose outputs were sent through band-limited amplifiers to a correlator. The theory of this experiment was developed later[78, 79] in a rather clumsy manner: the main results can be quite easily explained as follows.

The signal currents at the two photocells are proportional to the light intensities $I_1(t)$ and $I_2(t)$ falling on them, and therefore the correlation measured by Hanbury Brown and Twiss was proportional to $\langle I_1(t+\tau)\, I_2(t)\rangle$ which can be shown to obey the following relation, for a classical thermal source[89]:

$$\langle I_1(t+\tau)\, I_2(t)\rangle = I_1 I_2 [1 + \tfrac{1}{2}|\gamma_{12}(\tau)|^2]. \tag{7.26}$$

Alternatively, defining

$$\Delta I = I - \langle I\rangle, \tag{7.27}$$

we have

$$\begin{aligned}\langle \Delta I_1(t+\tau)\, \Delta I_2(t)\rangle &= \langle (I_1(t+\tau) - \langle I_1\rangle)(I_2(t) - \langle I_2\rangle)\rangle \\ &= \langle I_1(t+\tau)\, I_2(t)\rangle - \langle I_1\rangle\langle I_2\rangle,\end{aligned} \tag{7.28}$$

so that

$$\langle \Delta I_1(t+\tau)\, \Delta I_2(t)\rangle = \tfrac{1}{2}\langle I_1\rangle\langle I_2\rangle\, |\gamma_{12}(\tau)|^2. \tag{7.29}$$

The foregoing discussion does not describe the experimental situation fully. In fact, shot noise from the photocurrent is not taken into account. Moreover, although $I(t)$ changes slowly compared with the electric field, its fluctuations are too rapid for the electronic correlator to follow. In practice, the signals to be correlated in the Hanbury Brown and Twiss experiment are derived from band-limited electronic amplifiers, characterised by a certain frequency response $B(\nu)$. So, if we indicate with $S(t)$ the output of the linear filter, we have

$$S(t) = \alpha \int_0^\infty I(t-t')\, b(t')\, \mathrm{d}t', \tag{7.30}$$

where $b(t)$ is the response of the system to a short impulse at $t = 0$ and is the Fourier transform of $B(\nu)$. The measured correlation is therefore

$$\begin{aligned}\langle S_1(t)\, S_2(t)\rangle &= \alpha_1 \alpha_2 \iint_0^\infty \langle I_1(t-t')\, I_2(t-t'')\, b(t')\, b(t'')\, \mathrm{d}t'\, \mathrm{d}t'' \\ &= \alpha_1 \alpha_2 \iint_0^\infty \langle I_1(t+t''-t')\, I_2(t)\rangle\, b(t')\, b(t'')\, \mathrm{d}t'\, \mathrm{d}t'',\end{aligned} \tag{7.31}$$

and

$$\langle \Delta S_1(t)\, \Delta S_2(t)\rangle = \tfrac{1}{2}\alpha_1 \alpha_2 \langle I_1\rangle \langle I_2\rangle \,|\gamma_{12}(0)|^2 \times \iint |\gamma_{11}(t'' - t')|^2\, b(t')\, b(t'')\, \mathrm{d}t'\, \mathrm{d}t''$$

$$= \tfrac{1}{2}\alpha_1 \alpha_2 \langle I_1\rangle \langle I_2\rangle \,|\gamma_{12}(0)|^2 \times \int_0^\infty \int_{-t'}^\infty |\gamma_{11}(t''')|^2\, b(t')\, b(t' + t''')\, \mathrm{d}t'''\, \mathrm{d}t', \qquad (7.32)$$

where use has been made of the relation[90]

$$\gamma_{12}(\tau) = \gamma_{12}(0)\, \gamma_{11}(\tau). \qquad (7.33)$$

Now $|\gamma_{11}(\tau)|$ vanishes for τ much in excess of $1/\Delta\nu$. On the other hand, if the highest frequency passed by the electrical filter is still small compared with the bandwidth $\Delta\nu$ of the light, as is normally the case, then $b(t)$ is nearly constant over intervals of order $1/\Delta\nu$. Hence, for those values of t''' for which the integrand does not vanish, $b(t' + t''') \simeq b(t')$. Therefore

$$\langle \Delta S_1(t)\, \Delta S_2(t)\rangle = \tfrac{1}{2}\alpha_1\alpha_2 \langle I_1\rangle \langle I_2\rangle \,|\gamma_{12}(0)|^2\, \xi(\infty) \int_0^\infty b^2(t')\, \mathrm{d}t', \qquad (7.34)$$

where

$$\xi(\infty) = \int_{-\infty}^{+\infty} |\gamma_{11}(\tau)|^2\, \mathrm{d}\tau. \qquad (7.35)$$

Provided $\langle S_1\rangle$ and $\langle S_2\rangle$ are not zero this can be put into the form

$$\langle \Delta S_1(t)\, \Delta S_2(t)\rangle = \tfrac{1}{2}\langle S_1\rangle \langle S_2\rangle\, [\xi(\infty)/T']\, |\gamma_{12}(0)|^2, \qquad (7.36)$$

where

$$\left[\int_0^\infty b(t')\, \mathrm{d}t'\right]^2 \Big/ \left[\int_0^\infty b^2(t')\, \mathrm{d}t'\right] = T' \qquad (7.37)$$

is a rough measure of the time spread of the impulse response function $b(t)$, and therefore has the nature of a *resolving time* T'.

If noise is taken into account, the normalised signal correlation coefficient ρ can be written as

$$\rho = [\langle \Delta S_1(t)\, \Delta S_2(t)\rangle]\, [\langle \Delta S_1\rangle^2 \langle \Delta S_2\rangle^2]^{-1/2}$$

$$= [\tfrac{1}{2}\alpha \bar{I} \xi(\infty)]\, [1 + \tfrac{1}{2}\alpha \bar{I} \xi(\infty)]^{-1}\, |\gamma_{12}(0)|^2. \qquad (7.38)$$

The factor $\frac{1}{2}\alpha\bar{I}\xi(\infty)$, which gives half the average number of photoelectrons emitted in a time equal to the coherence time by a coherently illuminated photocathode, is called the degeneracy parameter δ[91].

For all classical light beams δ is very much less than unity. The result of the experiment can now be understood by looking at the meaning of $\gamma_{12}(0)$. This correlation function is defined as

$$\gamma_{12}(0) = \langle V(P_1, t)\, V^*(P_2, t)\rangle \,[\langle V(P_1, t)\, V^*(P_1, t)\rangle]^{-1/2}\,[\langle V(P_2, t)\, V^*(P_2, t)\rangle]^{-1/2}, \tag{7.39}$$

and gives the correlation of the disturbance at two different points at equal times. For a circular thermal source (such as a star) it had already been calculated by Hopkins[64] and is given in equation (7.23). Its square modulus is shown in the full line of figure 7.2. By varying the separation d between the two photomultipliers, the function $|\gamma_{12}(0)|^2$ is measured and therefore the parameters of the source (its angular magnitude) can also be measured. The result of the Hanbury Brown and Twiss experiment is shown in figure 7.2[79].

The result of this experiment confirmed the operational principle of an intensity interferometer as shown in figure 7.3. When using it for observing a star, the correlation will decrease with increasing baseline in a manner similar to that shown in figure 7.2 and then a measurement of the curvature will give the angular magnitude of the star.

The difference between such an interferometer and the conventional stellar Michelson interferometer is that the intensity interferometer measures the

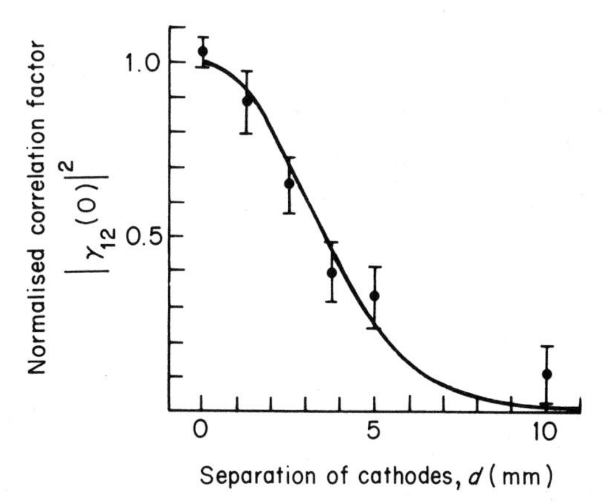

Figure 7.2 The experimental and theoretical values of the normalised correlation factor $|\gamma_{12}(0)|^2$ for different values of separation between the photocathodes. The full line is the theoretical curve and the experimental results are plotted as points with their associated probable errors[79].

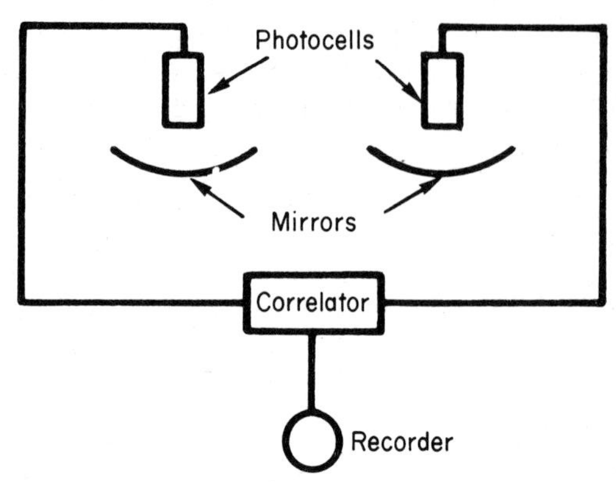

Figure 7.3 The Hanbury Brown and Twiss intensity interferometer at optical wavelengths[76]. Reprinted by permission from *Nature* **177**, 27; copyright © 1956 Macmillan Journals Ltd.

square of the modulus of the complex degree of coherence, while the Michelson one gives the complex degree of coherence itself, and so the phase of this complex function is lost. Roughly speaking, this means that one cannot reconstruct the angular distribution across an asymmetrical source without ambiguity; for example, when observing a double star with two unequal components, one cannot tell which star is on the left and which is on the right.

The first demonstration of time coincidence between photons was mentioned earlier (in Hanbury Brown's recollections). To obtain an adequate signal-to-noise ratio in a coincidence-counting experiment in which individual photons can be counted, an intense and narrow-band source of light is needed. Hanbury Brown and Twiss thought of using a mercury isotope lamp and found this type of lamp in Sydney, where the experiment was then performed in 1957 by Twiss and Little[87]. They used an electrode-less radio-frequency discharge in mercury-198 vapour as a source of light; 1P21 phototubes; and a coincidence counter with a resolving time of 3.5×10^{-9} s. The arrangement was similar to the one shown in figure 7.1. Both phototubes were mounted on movable slides, so that, as seen by the source, they could be optically superimposed or separated by a distance of 5 mm transverse to the line of sight; at this separation the pinhole source was completely resolved, so that the incident light beams were uncorrelated.

They compared the coincidences between photons arriving at these two phototubes with the coherent illumination (when the two phototubes were optically superposed) and the incoherent illumination (when the two phototubes were displaced). In a test lasting 8 h they found that, with coherent illumination, the number of coincidences was increased by $1.93 \pm 0.17\%$, which was in satisfactory agreement with the theoretical estimate of 2.07%.

Subsequently the correlation between photons was confirmed, using coincidence counters, by Rebka and Pound[92], and others[93].

The correlation observed between the photocurrent fluctuations of two photoelectric detectors immediately rules out the possibility that these currents consist of statistically independent pulses corresponding to the arrival of statistically independent photons at the photocathodes. Purcell[94] in 1956 was the first to point out that the effects observed could be explained quantitatively in terms of the quantum statistical behaviour of the photons; and these considerations were later followed up and extended by several authors[95]. Purcell began by considering one beam of light falling onto one photomultiplier and examined the statistical fluctuations in the counting rate. By assuming that the probability that a photoelectron is ejected in time $\mathrm{d}t$ can be written as $\alpha P \mathrm{d}t$, where α is a constant and P is the square of the electric field in the light, he derived the mean number of photoelectrons ($\bar{n}_T$) which will be counted in a time interval T as being

$$\bar{n}_T = \alpha \bar{P} T, \tag{7.40}$$

and showed that the variance,

$$\overline{(\Delta n)^2} = \overline{n^2} - \bar{n}^2, \tag{7.41}$$

satisfies the relation

$$\overline{(\Delta n)^2} = \bar{n}\,[1 + \bar{n}(\tau_0/T)], \tag{7.42}$$

where

$$\tau_0 = \int_{-\infty}^{+\infty} |\gamma_{11}(\tau)|^2\, \mathrm{d}\tau. \tag{7.43}$$

By assuming a quasi-monochromatic light with a Lorentzian lineshape of full width $\Delta\nu$ at half intensity

$$\tau_0 \propto 1/\Delta\nu, \tag{7.44}$$

so that

$$\overline{(\Delta n)^2} = \bar{n}\,[1 + (\bar{n}/\Delta\nu T)]. \tag{7.45}$$

This result shows that the fluctuations in the output current of a photoelectric detector are greater than the expected value ($\overline{\Delta n^2} = \bar{n}$) in a simple random stream. He adds:

> If one insists on representing photons by wave packets and demands an explanation in those terms of the extra fluctuations, such an explanation can be given....Think, then of a stream of wave packets, each about $c/\Delta\nu$ long, in a random sequence. There is a certain probability that two such trains accidentally overlap. When this occurs they interfere and one may find (to speak rather loosely) four photons, or none, or something in

between as a result. It is proper to speak of interference in this situation because the conditions of the experiment are just such as will ensure that these photons are in the same quantum state. To such interference one may ascribe the 'abnormal' density fluctuations in any assemblage of bosons.

Turning now to the split-beam experiment of Hanbury Brown and Twiss, let n_1 be the number of counts of one photomultiplier in an interval T, and let n_2 be the number of counts in the other in the same interval. As regards the fluctuations in n_1 alone, from interval to interval we face the situation already analysed, except that we shall now assume both polarisations present, which changes equation (7.42) into

$$\overline{(\Delta n_1^2)} = \bar{n}_1[1 + \tfrac{1}{2}\bar{n}_1(\tau_0/T)]. \tag{7.46}$$

A similar relation holds for n_2. Now, if we were to connect the two photomultiplier outputs together, we would revert to a single-channel experiment with count $n = n_1 + n_2$, or

$$\overline{(\Delta n^2)} = \bar{n}[1 + \tfrac{1}{2}\bar{n}(\tau_0/T)]. \tag{7.47}$$

But

$$\overline{(\Delta n^2)} = \overline{(\Delta n_1 + \Delta n_2)^2} = \bar{n}_1[1 + \tfrac{1}{2}\bar{n}_1(\tau_0/T)] + \bar{n}_2[1 + \tfrac{1}{2}\bar{n}_2(\tau_0/T)]$$
$$+ 2\Delta n_1 \Delta n_2. \tag{7.48}$$

From equations (7.47) and (7.48) it follows that

$$\overline{\Delta n_1 \Delta n_2} = \tfrac{1}{2}\bar{n}_1^2(\tau_0/T), \tag{7.49}$$

which is the positive cross-correlation effect of Hanbury Brown and Twiss. It was therefore demonstrated that the fluctuations in the counts recorded by a photoelectric detector which is illuminated by light from a thermal source are those to be expected from a boson assembly.

In 1958 Mandel[96, 97] derived classically the relation between the statistics of the photons of the field and the statistics of photoelectrons emitted by a photodetector. By developing Purcell's idea[94], Mandel began by saying[96]:

> We shall now associate photons with the Gaussian random wave $y(t)$, by defining a probability that a photoelectron is ejected in a short time interval between t and $t + \mathrm{d}t$. If we consider first-order transitions only significant, in which one photon gives rise to one photoelectron, then this probability will be given by $\alpha P(t)\,\mathrm{d}t$, where α is the quantum sensitivity of the photoelectric detector, assumed constant over the narrow frequency range $\Delta\nu_0$. The observable $P(t)$ provides the only link between the wave and the particle description of the beam.

The fluctuations of the number of particles n_T therefore have two causes. There are first of all the fluctuations of the wave intensity $P(t)$, determined by the spectral lineshape and there is the stochastic association of particles with the wave intensity. This twofold source of the fluctuations results in the departure from classical statistics, as we shall show[89].

He then wrote $p_n(t, T)$ for the probability that n photoelectrons are ejected in the interval between t and $t + T$. In particular, from the definition

$$p_1(t, T) = \alpha P(t)\, \mathrm{d}t, \tag{7.50}$$

and the expected value of n in the interval t to $t + T$ is

$$\alpha \int_t^{t+T} P(t')\, \mathrm{d}t'. \tag{7.51}$$

We may therefore expect from first principles to have

$$p_n(T) = \overline{p_n(t, T)} = (1/n!)\overline{\left[\alpha \int_t^{t+T} P(t')\, \mathrm{d}t'\right]^n \exp\left[-\alpha \int_t^{t+T} P(t')\, \mathrm{d}t'\right]}, \tag{7.52}$$

where the bar indicates time averaging.

The two papers[96, 97] are concerned with the calculation, on purely classical grounds, of $\bar{n}$, $\overline{n^2}$, and the problem of determining the statistics $p_n(T)$ for different values of T with respect to the coherence time τ of the light ($\tau \simeq 1/\Delta\nu_0$). In particular, when $\Delta\nu_0 T \ll 1$, $P(t)$ does not vary much in the time T, and it follows that

$$\alpha \int_t^{t+T} P(t')\, \mathrm{d}t' \simeq \alpha P(t)\, T. \tag{7.53}$$

Therefore

$$\overline{p_n(t, T)} = (1/n!)\overline{(\alpha PT)^n \exp(-\alpha PT)}. \tag{7.54}$$

If the probability distribution P is known, $p_n(t, T)$ can be evaluated by averaging over the ensemble. Since $y(t)$ is a narrow-band Gaussian random variable, the local average is

$$\langle y^2(t) \rangle = P(t) = \tfrac{1}{2} w^2(t), \tag{7.55}$$

where $w(t)$ is the envelope of $y(t)$. Now, the probability density of $w(t)$ was shown by Rice[98] to be of the form

$$(w/\bar{P}) \exp(-w^2/2\bar{P}).$$

Hence

$$P'(P)\,\mathrm{d}P = (1/\bar{P})\exp[-(P/\bar{P})]\,\mathrm{d}P, \tag{7.56}$$

and

$$\overline{p_n(t, T)} = \{(1 + \alpha\bar{P}T)[1 + (1/\alpha\bar{P}T)]^n\}^{-1} \tag{7.57}$$

with

$$\overline{\Delta n^2} = \bar{n}(1 + \bar{n}). \tag{7.58}$$

This is the well known formula for the fluctuations of the occupation numbers of a *single cell* in phase space for an assembly of bosons. Mandel[96] says:

> The photoelectrons in the interval T therefore obey 'pure' Bose-Einstein statistics. The reason for this can be seen at once if we examine the size of the elementary cell in phase space. In the direction of the beam this extends over a distance $c\Delta t$, where $\Delta t \sim 1/\Delta\nu_0$. Thus, the photons in an interval $T \ll 1/\Delta\nu_0$ as above, i.e. much shorter than the so-called coherence time of the light..., occupy the same cell in phase space. By the uncertainty principle they are therefore intrinsically indistinguishable and n obeys pure Bose-Einstein statistics.

When $\Delta\nu_0 T \gg 1$, Mandel[96] still derived

$$\overline{\Delta n^2} = \bar{n}(1 + k\bar{n}/\Delta\nu_0 T), \tag{7.59}$$

where k is a number depending on the spectral density; this is the relation obtained by Purcell.

In this case we are now dealing with a volume of phase space containing roughly $\Delta\nu_0 T = s$ cells. The mean number of photons per cell is therefore $\bar{n}/s = \bar{m}$, and equation (7.59) can be written

$$\overline{\Delta n^2} = \bar{n}(1 + k\bar{m}) = \bar{n}(1 + ks\overline{m^2}/\bar{n}). \tag{7.60}$$

This is the expression in the conventional form for the density fluctuations in a larger volume of phase space which was first found by Furth[42, 43].

These considerations could be perfected considering the degeneracy parameter δ. L Mandel in 1961[99] illustrated this point quite well. δ is the average number of photons in the light beam which are to be found in the same quantum state, or in the same cell of phase space. The correlation between the fluctuations of the counts recorded by two photomultipliers illuminated by partially coherent light from a thermal source is proportional to δ. The degeneracy of black-body radiation in an enclosure at a temperature T was first shown by Einstein[100] to be

$$\delta = [\exp(h\nu/kT) - 1]^{-1} \tag{7.61}$$

at a frequency ν. It was not immediately obvious that this relation was also valid for a light beam far removed from its equilibrium source, but Mandel showed that this was the case. These results on equilibrium photon statistics received full support some years later in 1964, when Mandel, Sudarshan and Wolf[101] derived, on a semiclassical basis, that the probability $p(t)\,\Delta t$ of photoemission of an electron is proportional to the classical measure of the instantaneous light intensity

$$p(t)\,\Delta t = \alpha I(t)\,\Delta t, \tag{7.62}$$

which is the same as equation (7.50).

7.7 The quantum theory of coherence

In the meantime a revolution had taken place in coherence theory due to a few papers written by R J Glauber.

The main problem, as we have seen, was to understand the essence of the Hanbury Brown and Twiss experiment. The revolution started quietly with a letter to *Physical Review Letters*[102]. It begins

> In 1956 Hanbury Brown and Twiss[102] reported that the photons of a light beam of narrow spectral width have a tendency to arrive in correlated pairs. We have developed general quantum mechanical methods for the investigation of such correlation effects and shall present here results for the distribution of the number of photons counted in an incoherent beam. The fact that photon correlations are enhanced by narrowing the spectral bandwidth has led to a prediction (by Mandel and Wolf[102]) of large-scale correlations to be observed in the beam of an optical maser. We shall indicate that this prediction is misleading and follows from an inappropriate model of the maser beam.

In making this last statement Glauber was rather sharp, and this issue raised a vigorous controversy between Glauber and the Rochester group which lasted several years, as we shall see later.

In his paper Glauber then sketches a quantum mechanical representation of the field, introducing the *coherent states* representation and defining the density operator appropriate to the problem. He then discusses the photon correlations and shows that coherent states of the field lead to no photoionisation correlations whatsoever. A correlation between photons only appears when either incoherent mixtures or superpositions of the coherent states are present. Although he has no precise idea of the density operator for an actual laser, he then supposes it is more likely to be of the kind produced by a product of coherent states.

In two subsequent, very beautiful, papers in *Physical Review* of the same year[103, 104], Glauber fully develops the quantum theory of optical coherence. He takes as his starting point the separation of the electric field operator $\mathbf{E}(\mathbf{r}t)$ into its positive and negative frequency parts:

$$\mathbf{E}(\mathbf{r}t) = \mathbf{E}^{(+)}(\mathbf{r}t) + \mathbf{E}^{(-)}(\mathbf{r}t). \tag{7.63}$$

The positive part is associated with photon absorption and the negative part with photon emission. In particular, the positive frequency part, $\mathbf{E}^{(+)}(\mathbf{r}t)$, may be shown[110] to be a photon annihilation operator. Applied to an n-photon state it produces an $(n-1)$-photon state.

The Hermitian adjoint of it, $\mathbf{E}^{(-)}(\mathbf{r}t)$ creates a photon. Applied to an n-photon state it produces an $(n+1)$-photon state.

This decomposition of the electric field has a very close connection with classical treatment. Before second quantisation the positive frequency part $\mathbf{E}^{(+)}$ can indeed be identified with the classical analytical signal discussed above.

Glauber then shows that a detection process such as photoionisation, in which the field makes a transition from the initial state $|i\rangle$ to a final state $|f\rangle$ in which one photon has been absorbed, is described by the matrix element

$$\langle f|E^{(+)}(\mathbf{r}t)|i\rangle,$$

and the probability per unit time that a photon be absorbed by an ideal detector at point r at time t is proportional to

$$\sum_f |\langle f|E^{(+)}(\mathbf{r}t)|i\rangle|^2 = \langle i|E^{(-)}(\mathbf{r}t)\,E^{(+)}(\mathbf{r}t)|i\rangle. \tag{7.64}$$

Recording photon intensities with a single detector does not exhaust the measurements one can make on the field, though it does characterise, in principle, virtually all the classic experiments of optics. A second type of measurement consists of the use of two detectors situated at different points, r and r' to detect photon coincidences or, more generally, delayed coincidences. The field matrix element for such transitions takes the form

$$\langle f|E^{(+)}(\mathbf{r}'t')\,E^{(+)}(\mathbf{r}t)|i\rangle,$$

and the total rate at which such transitions occur is proportional to

$$\sum_f |\langle f|E^{(+)}(\mathbf{r}'t')\,E^{(+)}(\mathbf{r}t)|i\rangle|^2$$
$$= \langle i|E^{(-)}(\mathbf{r}t)\,E^{(-)}(\mathbf{r}'t')\,E^{(+)}(\mathbf{r}'t')\,E^{(+)}(\mathbf{r}t)|i\rangle. \tag{7.65}$$

Such a rate is the one connected with experiments of the type performed by Hanbury Brown and Twiss.

More elaborate experiments can also be considered, in which detection of n-field delayed coincidences of photons for arbitrary n is considered. The total rate per unit time for such coincidences will be proportional to

$$\langle i|E^{(-)}(\mathbf{r}_1 t_1)\ldots E^{(-)}(\mathbf{r}_n t_n)\, E^{(+)}(\mathbf{r}_n t_n)\ldots E^{(+)}(\mathbf{r}_1 t_1)|i\rangle.$$

At this point Glauber observes that the electromagnetic field may be regarded as a dynamical system with an infinite number of degrees of freedom. Our knowledge of the condition of such a system is, in practice, virtually never so complete or so precise as to justify in its description the use of a particular quantum state $|\rangle$.

We have therefore to consider averages over the distributions of the unknown parameters. Such averages can be constructed quantum mechanically by making recourse to an Hermitian operator known as a *density operator* ρ which is constructed as an average over the uncontrollable parameters of an expression which is bilinear in the state vector[105]. If $|\rangle$ is a precisely defined state of the field corresponding to a particular set of random parameters

$$\rho = \{\,|\rangle\langle|\,\}_{av}.$$

The average of an observable θ in the quantum state $|\rangle$, over the randomly prepared states is the quantity of interest in experiments and it is given by

$$\{\langle|\theta\,|\rangle\}_{av} = \mathrm{Tr}\{\,\rho\theta\,\},$$

where the symbol Tr stands for the trace, or sum, of the diagonal matrix elements.

The average counting rate of an ideal photodetector, for example, is therefore proportional to

$$G(\mathbf{r}t, \mathbf{r}t) = \mathrm{Tr}\{\,\rho E^{(-)}(\mathbf{r}t)\, E^{(+)}(\mathbf{r}t)\}. \tag{7.66}$$

Glauber now introduces the set of functions

$$\begin{aligned} &G^{(n)}(x_1 \ldots x_n, x_{n+1}, \ldots, x_{2n}) \\ &\quad = \mathrm{Tr}\{\rho E^{(-)}(x_1)\ldots E^{(-)}(x_n)\, E^{(+)}(x_{n+1})\ldots E^{(+)}(x_{2n})\} \end{aligned} \tag{7.67}$$

which represent the correlation functions of the field at different space-time points, $x_j = r_j t_j$ being the $G^{(n)}$ function necessary to discuss an n-photon coincidence experiment.

By introducing the normalised forms

$$g^{(n)}(x_1 \ldots x_{2n}) = G^{(n)}(x_1 \ldots x_{2n}) \Big/ \prod_{j=1}^{2n} \{G^{(1)}(x_j x_j)\}^2, \tag{7.68}$$

Glauber now extends the notion of coherence, defining different orders of coherence – first-, second-, third-order of coherence and so on – according to which the sequence

$$|g^{(n)}(x_1 \ldots x_{2n})| = 1, \tag{7.69}$$

for every n value, i.e. for $n = 1, n = 2, n = 3$ and so on.

First-order coherence, or

$$|g^{(1)}(x_1 x_2)| = 1, \tag{7.70}$$

is what is required for classical optics coherence experiments. The introduction of ensemble averages in place of time averages as used previously extended the notion of coherence to non-stationary fields.

In the next paper[104] Glauber retraces the full treatment of quantisation of the electromagnetic field and introduces the notion of *coherent states*, and the *P representation*[106], surmising that the laser could be in a coherent state. The starting point of the standard technique of quantisation of electromagnetic radiation *in vacuo* is Maxwell's equations written in the absence of charges and currents. The vector potential $\mathbf{A}(\mathbf{r}t)$ is expanded in terms of a complete set of real orthogonal mode functions $\mathbf{u}_l(\mathbf{r})$ with real coefficients $q_l(t)$

$$\mathbf{A}(\mathbf{r}t) = \sum_l q_l(t)\, \mathbf{u}_l(\mathbf{r}), \tag{7.71}$$

and it is shown that the total energy of the field

$$H_0 = (1/8\pi) \int_{\text{cavity}} (\mathbf{E}^2 + \mathbf{H}^2)\, \mathrm{d}\mathbf{r} \tag{7.72}$$

can be expressed as

$$H_0 = \tfrac{1}{2} \sum_l [p_l^2(t) + \omega_l^2 q_l^2(t)] \tag{7.73}$$

where

$$p_l(t) = \mathrm{d}q_l/\mathrm{d}t. \tag{7.74}$$

In this way, the electromagnetic field is described in terms of a set of independent couples of conjugate variables, q_l and p_l, relative to a set of independent harmonic oscillators. The quantisation of the electromagnetic field is now achieved by regarding the q_l and p_l as Hermitian operators obeying the commutation relations

$$[p_l, p_m] = [q_l, q_m] = 0 \qquad [q_l, p_m] = \mathrm{i}\hbar\delta_{lm}, \tag{7.75}$$

according to a basic postulate of quantum mechanics. The q_l and p_l will be explicitly time independent or not, according to whether or not the Schrödinger picture is used. The standard procedure of quantisation of the harmonic oscillator consists, then, in the introduction of a pair of non-Hermitian operators a_l^+ and a_l by means of the equations

$$q_l = (\hbar/2\omega_l)^{1/2}[a_l^+ + a_l], \tag{7.76}$$

$$p_l = \mathrm{i}(\hbar\omega_l/2)^{1/2}[a_l^+ - a_l]. \tag{7.77}$$

It can easily be seen from equations (7.76) and (7.77) that a_l and a_l^+ are Hermitian conjugate operators, while equation (7.75) shows that they obey the commutation relations

$$[a_l, a_m^+] = \delta_{lm} \qquad [a_l, a_m] = [a_l^+, a_m^+] = 0. \tag{7.78}$$

The Hamiltonian of the system immediately follows from equations (7.73), (7.76) and (7.78) as

$$H_0 = \sum_l H_l = \sum_l \hbar\omega_l(a_l^+ a_l + a_l a_l^+). \tag{7.79}$$

If we now choose to use the Heisenberg picture, which is the most appropriate one if a comparison between classical and quantum treatment is to be made, the time evolution of $a_l(t)$ and $a_l^+(t)$ is determined by the Heisenberg equations of motion

$$\mathrm{i}\hbar\,(\mathrm{d}a_l(t)/\mathrm{d}t) = [a_l(t), H_0] = \hbar\omega_l a_l(t) \tag{7.80}$$

$$\mathrm{i}\hbar\,[\mathrm{d}a_l^+(t)/\mathrm{d}t] = [a_l^+(t), H_0] = -\hbar\omega_l a_l^+(t), \tag{7.81}$$

having taken into account equations (7.78) and (7.79). One has, according to equations (7.80) and (7.81)

$$a_l(t) = a_l \exp(-\mathrm{i}\omega_l t) \qquad a_l^+(t) = a_l^+ \exp(\mathrm{i}\omega_l t) \tag{7.82}$$

where $a_l = a_l(0)$ and $a_l^+ = a_l^+(0)$ are, from now on, these operators in the Schrödinger picture.

The eigenvalues n_l of the operator associated with the eigenvalue equation

$$a_l^+ a_l |n_l\rangle = n_l |n_l\rangle, \tag{7.83}$$

are furnished by all the non-negative integers, so that n_l can be interpreted as the number of energy quanta in the mode l.

This allows us to give $a_l^+ a_l$ the meaning of a numerical operator. Furthermore, a_l^+ and a_l are usually termed 'creation' and 'annihilation' operators, since they can be shown, respectively, to increase and lower by one the number of quanta, when they are operating on the eigenstates of $a_l^+ a_l$, according to

$$a_l^+ |n_l\rangle = (n_l + 1)^{1/2} |n_l + 1\rangle \tag{7.84}$$

$$a_l |n_l\rangle = (n_l)^{1/2} |n_l - 1\rangle. \tag{7.85}$$

The preceding considerations summarise the properties of the vector potential operator, which, according to equations (7.71) and (7.76) can be written in the Heisenberg picture as

$$\mathbf{A}(\mathbf{r}t) = \sum_{l} (\hbar/2\omega_l)^{1/2} [a_l \exp(-\mathrm{i}\omega_l t) + a_l^{\dagger} \exp(\mathrm{i}\omega_l t)]\, \mathbf{u}_l(\mathbf{r}). \quad (7.86)$$

The electromagnetic radiation field is then completely specified, from a quantum mechanical point of view, once the initial state of the system is assigned in terms of the $|n_l\rangle$.

The coherent states $|\alpha\rangle$ are now defined[107], for a single-mode case, in terms of the eigenstates $|n\rangle$ of the numerical operators (see equation (7.83)) as

$$|\alpha\rangle = \exp(-|\alpha|^2/2) \sum_{n=0}^{\infty} [\alpha^n/(n!)^{1/2}]\,|n\rangle, \quad (7.87)$$

and verify the eigenvalue equation

$$a|\alpha\rangle = \alpha|\alpha\rangle, \quad (7.88)$$

for any complex number α. They are not mutually orthogonal and form an over-complete set, in the sense that every $|\alpha\rangle$ can be expanded as a linear combination of the others. The generalisation to the multi-mode case is easily performed by the introduction of the coherent state $|\{\alpha_l\}\rangle$, defined as the product of the coherent states $|\alpha_l\rangle_l$ relative to each mode

$$|\{\alpha_l\}\rangle = \prod_{l} |\alpha_l\rangle_l. \quad (7.89)$$

It is then easy to show that they are eigenkets of the positive frequency part of the electric field operator

$$\mathbf{E}^{(+)}(\mathbf{r}t)\,|\{\alpha_l\}\rangle = \mathscr{E}_{\{\alpha_l\}}(\mathbf{r}t)\,|\{\alpha_l\}\rangle, \quad (7.90)$$

with complex eigenvalues $\mathscr{E}_{\{\alpha_l\}}(\mathbf{r}t)$ given by

$$\mathscr{E}\{\alpha_l\}(\mathbf{r}t) = \mathrm{i} \sum_{l} (h\omega_l/L^3)^{1/2} \alpha_l \exp[\mathrm{i}(\mathbf{k}_l \cdot \mathbf{r} - \omega_l t)], \quad (7.91)$$

and, consequently, that

$$\langle\{\alpha_l\}|\,\mathbf{E}^{(-)}(\mathbf{r}t) = \mathscr{E}^{*}_{\{\alpha_l\}}(\mathbf{r}t)\langle\{\alpha_l\}|. \quad (7.92)$$

According to equations (7.90) and (7.92), the correlation function $G^{(n)}$ defined by

$$G^{(n)}(r_1 t_1, \ldots, r_n t_n, r_n t_n, \ldots, r_1 t_1)$$
$$= \{\langle i|E^{(-)}(r_1 t_1) \ldots E^{(-)}(r_n t_n)\, E^{(+)}(r_n t_n) \ldots E^{(+)}(r_1 t_1)|i\rangle\}, \quad (7.93)$$

averaged over $|i\rangle$, assumes, for a coherent state, the particularly simple form

$$G^{(n)}(\mathbf{r}_1t_1,\ldots,\mathbf{r}_nt_n,\mathbf{r}_nt_n,\ldots,\mathbf{r}_1t_1)$$
$$=\langle\{\alpha_l\}|\,\mathbf{E}^{(-)}(\mathbf{r}_1t_1)\ldots\mathbf{E}^{(-)}(\mathbf{r}_nt_n)\,E^{(+)}(\mathbf{r}_nt_n)\ldots E^{(+)}(\mathbf{r}_1t_1)|\{\alpha_l\}\rangle$$
$$=\mathscr{E}^*_{\{\alpha_l\}}(\mathbf{r}_1t_1)\ldots\mathscr{E}^*_{\{\alpha_l\}}(\mathbf{r}_nt_n)\,\mathscr{E}_{\{\alpha_l\}}(\mathbf{r}_nt_n)\ldots\mathscr{E}_{\{\alpha_l\}}(\mathbf{r}_1t_1). \tag{7.94}$$

Thus in this case, for which no statistical uncertainty is present to our knowledge of the state of the system, the quantum mechanical expression for $G^{(n)}$ is equivalent to the classical expression given by

$$G^{(n)}(\mathbf{r}_1t_1,\ldots,\mathbf{r}_nt_n,\mathbf{r}_nt_n,\ldots,\mathbf{r}_1t_1)$$
$$=\langle\hat{E}^*(\mathbf{r}_1t_1)\ldots\hat{E}^*(\mathbf{r}_nt_n)\,\hat{E}(\mathbf{r}_nt_n)\ldots\hat{E}(\mathbf{r}_1t_1)\rangle, \tag{7.95}$$

for a prescribed electromagnetic field with analytic signal $\mathscr{E}_{\{\alpha_l\}}(\mathbf{r}t)$. This equivalence holds independent of the average number of photons associated with the state $|\{\alpha_l\}\rangle$, the ultimate difference between the quantum mechanical and classical cases lying in the fact that $|\{\alpha_l\}\rangle$ is an eigenstate of $E^{(+)}(\mathbf{r}t)$ and not of the electric field operator $E(\mathbf{r}t)=E^{(+)}(\mathbf{r}t)+E^{(-)}(\mathbf{r}t)$.

The preceding analogy was then extended to the situation in which one deals with a statistical mixture of coherent states, introducing the density matrix ρ given by

$$\rho=|\{\alpha_l\}\rangle\langle\{\alpha_l\}|. \tag{7.96}$$

More generally, we can consider the situation in which ρ is expressed as a linear combination of operators $|\{\alpha_l\}\rangle\langle\{\alpha_l\}|$ in the form

$$\rho=\int P(\{\alpha_l\})|\{\alpha_l\}\rangle\langle\{\alpha_l\}|\,\mathrm{d}^2\{\alpha_l\}, \tag{7.97}$$

with $\mathrm{d}^2\{\alpha_l\}\equiv\Pi_l\,\mathrm{d}(\mathrm{Re}\,\alpha_l)\,\mathrm{d}(\mathrm{Im}\,\alpha_l)$. This representation for the density operator was introduced by Glauber[104] and Sudarshan[106] and is known as *P representation*[108]. An example of this representation was given[104] for thermal (chaotic) occupation of the mode for Gaussian light.

In order to preserve the Hermitian and unitary character of the operator ρ, the P representation must be real and satisfy the normalisation condition

$$\int P(\{\alpha_l\})\,\mathrm{d}^2\{\alpha_l\}=1, \tag{7.98}$$

while it may take on negative values. The possibility of expressing the density matrix in terms of the P representation actually exists for a large class of physical situations and allows us to write the correlation function $G^{(n)}$ as

$$G^{(n)}=\mathrm{Tr}\,[\rho E^{(-)}(\mathbf{r}_1t_1)\ldots E^{(-)}(\mathbf{r}_nt_n)\,E^{(+)}(\mathbf{r}_nt_n)\ldots E^{(+)}(\mathbf{r}_1t_1)]$$
$$=\int P(\{\alpha_l\})|\mathscr{E}_{\{\alpha_l\}}(\mathbf{r}_1t_1)|^2\ldots|\mathscr{E}_{\{\alpha_l\}}(\mathbf{r}_nt_n)|^2\,\mathrm{d}^2\{\alpha_l\}. \tag{7.99}$$

This expression is formally equivalent to what would be written in a classical situation for which equation (7.95) is valid, bearing in mind the fact that one would deal with a non-negative weight function having the meaning of probability density.

These considerations make clear the important role played by the P representation, which, whenever it exists, enables us to perform, in a formally identical way, classical and quantum mechanical calculations relevant to the evaluation of the hierarchy of the correlation functions $G^{(n)}$. In any event, the $P(\{\alpha_l\})$ cannot be interpreted, even when it assumes only positive values, as a probability density of finding the system in a given state $|\{\alpha_l\}\rangle$, since the coherent states are not mutually orthogonal. This interpretation becomes practically valid in the classical limit of a large number of photons in which $P(\{\alpha_l\})$ tends to coincide with the corresponding probability density of finding a set of mode amplitudes $\{\alpha_l\}$, the structure of the coherent states being such that $\langle\{\alpha_l\}|\{\alpha_l'\}\rangle$ approaches zero in the classical limit for $\{\alpha_l\} \neq \{\alpha_l'\}$[104, 109].

Later, Glauber[110] fully discussed the detection process, showing that, if one considers a broadband photoelectric detector, the probability $p^{(1)}(t)$ of counting a photon is given by

$$p^{(1)}(t) = s\int_{t_0}^{t_1} G^{(1)}(rt', rt')\, \mathrm{d}t', \tag{7.100}$$

where the *sensitivity* s summarises the response of the detecting atomic system.

The expression given by equation (7.100) is fully quantum mechanical, since both the radiation and the detecting system have been quantised. In the classical limit for the radiation field one expects the quantum mechanical operators $E^{(+)}$ and $E^{(-)}$ to be, respectively, replaced by the analytic signal $\hat{E}$ and its complex conjugate. This was rigorously proved by Glauber[104], and it is in perfect agreement with the semiclassical treatment in which only the atomic system is quantised. In this situation, the quantum mechanical quantity $G^{(1)}$ can be approximated by the classical ensemble average

$$G^{(1)}(\mathbf{r}t, \mathbf{r}t) = \langle \hat{E}^*(\mathbf{r}t)\, \hat{E}(\mathbf{r}t)\rangle. \tag{7.101}$$

We now observe that, in a real counting device, $t - t_0$ contains a great number of oscillation periods of the radiation field, so that $p^{(1)}(t)$ is, in practice, an average of $G^{(1)}(\mathbf{r}t, \mathbf{r}t)$ over many periods. On the other hand, it can be easily shown[61] that $\hat{E}^*(\mathbf{r}t)\, \hat{E}(\mathbf{r}t)$ is, for quasi-monochromatic fields, proportional to the time average over a few periods of the instantaneous intensity $I(\mathbf{r}t)$; so that equation (7.100) is, in the classical limit, equivalent to

$$p^{(1)}(t) = s'\int_{t_0}^{t} \langle I(\mathbf{r}t')\rangle\, \mathrm{d}t', \tag{7.102}$$

with $s' = (8\pi/c)\, s$.

In the case of an actual detector containing a great number N of independent identical atoms, which are supposed to be uniformly illuminated by the radiation field, the average number of counts $\langle C(t)\rangle$ recorded in the time interval $t-t_0$ is given by

$$\langle C(t)\rangle = Np^{(1)}(t). \tag{7.103}$$

Equations (7.102) and (7.103) furnish a justification for the classical statement that the response of a photoelectric detector is proportional to the incident intensity. The quantum generalisation given by equation (7.100) shows that the relevant operator corresponding to the classical intensity is $E^{(-)}(\mathbf{r}t)\,E^{(+)}(\mathbf{r}t)$. This is a significant result since it provides an example in which the quantisation of a classical law is not *a priori* unique and requires a precise physical insight for determining the correct ordering of the operators involved.

Under the same assumptions which lead to equation (7.100), Glauber showed that the probability $p^{(n)}(t)$ of counting n photons in a time interval $t-t_0$ with n identical one-atom detectors placed at different positions $\mathbf{r}_i$ $(i=1,2,\ldots,n)$ is

$$p^{(n)}(t) = s^n \int_{t_0}^{t} \mathrm{d}t_1 \ldots \int_{t_0}^{t} \mathrm{d}t_n G^{(n)}(\mathbf{r}_1 t_1, \ldots, \mathbf{r}_n t_n,\ \mathbf{r}_n t_n, \ldots, \mathbf{r}_1, t_1). \tag{7.104}$$

This in turn implies the generalisation of equation (7.102) in the form

$$p^{(n)}(t) = s'^n \int_{t_0}^{t} \mathrm{d}t_1 \ldots \int_{t_0}^{t} \mathrm{d}t_n\ \langle I(\mathbf{r}_1 t_1)\ldots I(\mathbf{r}_n t_n)\rangle. \tag{7.105}$$

Glauber then considered the probability $p(m,t)$ of counting m photons in a given time interval 0–t (where m is any integer smaller than or equal to N), and showed[110] that a way of expressing $p(m,t)$ in terms of the statistical properties of the field can be made very elegantly by introducing a generating function $Q(\lambda,t)$ defined through the expectation value

$$Q(\lambda,t) = \mathrm{Tr}\,[\rho(1-\lambda)^{C(t)}] = \langle(1-\lambda)^{C(t)}\rangle, \tag{7.106}$$

where $C(t)$ is the operator number of photons registered in the time interval $(0,t)$. Accordingly, the meaning of $p(m,t)$ allows us to write

$$Q(\lambda,\ t) = \sum_{m=0}^{\infty} (1-\lambda)^m p(m,t), \tag{7.107}$$

from which it follows that

$$p(m,t) = [(-1)^m/m!][(\mathrm{d}^m/\mathrm{d}\lambda^m)\,Q(\lambda,t)]_{\lambda=1}. \tag{7.108}$$

Introducing now the factorial moments M_k, defined as

$$M_k = \langle C(C-1)\ldots(C-k+1)\rangle = \sum_{m=0}^{\infty} m(m-1)\ldots(m-k+1)\,p(m,t), \tag{7.109}$$

he showed that

$$M_k = \beta^k \int_0^t dt_1 \ldots \int_0^t dt_k \int_{V_c} d\mathbf{r}_1 \ldots \int_{V_c} d\mathbf{r}_k$$

$$\times\, G^{(k)}(\mathbf{r}_1 t_1, \ldots, \mathbf{r}_k t_k, \mathbf{r}_k t_k, \ldots, \mathbf{r}_1 t_1). \qquad (7.110)$$

In the same year, F Ghielmetti[111] showed that $p(n)$ may also be written as

$$p(n) = \int P(\{\alpha_k\})(U^n/n!) \exp(-U)\, d^2\{\alpha_k\},$$

where

$$U = \sum_k |\alpha_k|^2,$$

which may alternatively be cast in the form

$$p(n) = \int_0^\infty P(U')(U'^n/n!) \exp(-U')\, dU',$$

where

$$P(U') = \int P(\{\alpha_k\})\, \delta(U' - U)\, d^2\{\alpha_k\},$$

which can be compared with equation (7.54) as obtained by Mandel[96]. It is interesting to observe that somewhat similar results had already been obtained by Bothe[41].

Further insight into the properties of correlation functions and statistics was given by Mandel[112], who observed that the quantity

$$k_n = \int_{\delta V} \ldots \int \langle E^{(-)}(x_1) E^{(+)}(x_1) \ldots E^{(-)}(x_n) E^{(+)}(x_n)\rangle\, d^3x_1 \ldots d^3x_n$$

represents the nth moment of the number operator for photons in a volume δV. So, unordered products of creation and annihilation operators correspond to counting moments, while normal, ordered products of the kind of equation (7.93) are measures of the n-fold coincidence-counting rate for photoelectric detectors. Later Mandel[113] again considered the problem and showed that antinormally ordered operators of the form

$$G'^{(nn)}(x_1 \ldots x_n) = \langle E^{(+)}(x_1) E^{(-)}(x_1) \ldots E^{(+)}(x_n) E^{(-)}(x_n)\rangle$$

can be connected to photon counters functioning by stimulated emission rather than by absorption of photons, and discussed some of the properties of these correlations.

7.8 The discussion of the need for quantum optics

At first this beautiful construction of Glauber's was attacked on all sides. L Mandel and E Wolf immediately replied to the *Physical Review Letters* paper[102], and wrote[114]:

> In an interesting recent note published in this journal, Glauber[102] refers to one of our papers [Mandel and Wolf[102]] with a comment that a prediction made in that paper is misleading and follows from an inappropriate model of the maser beam. To the extent that we refer to the possible use of our results in connection with optical masers, our remarks were indeed misleading . . .

and they follow to clarify this point. They then added:

> There is also an implication in reference (1)[102] that the stochastic semiclassical description of light, initiated by Bothe[41] and Purcell[94] and developed further elsewhere[68, 96, 97, 114], ceases to be useful when the wave field cannot be represented classically by a Gaussian probability distribution. We do not believe this to be the case . . .

This was the start of a controversy which was to last several years. Indeed, E Wolf[115] and L Mandel[116], at the 3rd Quantum Electronics Conference, held in Paris in 1963, presented a classical treatment in which correlation functions in terms of the analytic signal were introduced, showing that this treatment was sufficient to explain all experimental results obtained until that moment, including the Hanbury Brown and Twiss experiment. No mention was made of Glauber's work. Glauber immediately reacted and, being concerned about the logical inconsistency of mixed treatments of classical and quantum mechanics, stated, in a comment on Wolf's paper, 'I think you should treat the problem quantum mechanically'.

This issue at once became the subject of the controversy. Wolf replied cautiously

> . . . Of course, one should try to formulate a full quantum mechanical treatment of coherence, but this may not be very easy to do. For many purposes the classical and semiclassical treatments are quite good approximations and in fact have been extremely successful in predicting the results of experiments.
>
> One should also bear in mind that the classical theory arose from an attempt to understand certain types of phenomena with light from thermal sources. Of course, as new problems arise, the theory has to be extended and this is precisely what is now being done with the help of

higher order correlation functions. But my guess is that for maser light classical theories will be even more useful than for thermal light.

In fact, however, a considerable effort was made by the Rochester group to prove that everything could be treated perfectly in a fully classical way. A point in favour of this was the propounding of the so-called 'optical equivalence theorem' by E C G Sudarshan[106], in which it was shown that every time the ρ matrix can be reduced to a diagonal representation one is able to formulate the quantum theory of optical coherence in a language formally equivalent in all essential respects to the language of classical theory, as expressed in terms of analytic signals. However, Cahill[117] showed that, whenever there is no upper bound to the number of quanta present, the P representation is missing.

With time, the quarrel died down and, in their beautiful review paper on coherence, Mandel and Wolf[60] also used Glauber's quantum treatment. The elegant construction of the quantum theory of coherence was, from that moment on, accepted unconditionally.

Meanwhile the meaning of quantum correlation functions was fully investigated by Titulaer and Glauber[118]. The first-order coherence condition implies maximum fringe contrast in interference patterns. In this paper they investigate the mathematical consequences of assuming the condition for maximum fringe contrast. They showed that this condition in turn implies factorisation of the first-order correlation function. They were then able to show that all of the higher-order correlation functions factorise into forms similar to those required for full coherence, but differing from them through the inclusion of a sequence of constant numerical factors. These coefficients furnish a convenient description of the higher-order coherence properties of the field.

An interesting property of fields with positive-defined P functions was then shown, i.e. that the combination of first- and second-order coherence implies coherence for all orders.

For some time only chaotic fields, coherent laser fields and mixtures of these were considered in the discussion. It was only later that the possibility of other kinds of fields was demonstrated. The first ones to be studied were types of super-Gaussian fields[119] for which the coherence properties were different from both the Gaussian and the coherent fields.

In addition, the first practical application of the theory to some real problem was only given in 1967[120] with the demonstration that, in a scattering experiment where a fully coherent field is used as a probe (plane wave), the correlation functions of the scattered field are reminiscent of the correlation functions of the fluctuations in the medium. This result, applied to fluids, may enable the testing of hypotheses on the origin of fluctuations[127]. Moreover, there were cases where the previous considerations had led to interesting new kinds of measurements. The first of these cases was shown to exist in turbulence[122].

7.9 Experimental studies of the statistical properties of light

The theoretical papers on coherence and statistics which we looked at in § 7.7 led to several predictions concerning the statistical properties of fields, which are summarised here[110].

7.9.1 *Polarised thermal light*

In the case of polarised thermal light the wave field is described by a scalar random process $V(t)$ in the form of an analytic signal which obeys a Gaussian distribution.

The instantaneous intensity $I(t)$ is given by

$$I(t) = V^*(t)\, V(t), \tag{7.111}$$

and the probability density of I is an exponential function

$$p(I) = (1/\langle I\rangle) \exp\{-I/\langle I\rangle\}, \tag{7.112}$$

where

$$\langle I\rangle = \lim_{T\to\infty} (1/T) \int_0^T I(t')\, \mathrm{d}t' \tag{7.113}$$

is the mean intensity.

The integrated light intensity $U = \int_0^T I(t')\, \mathrm{d}t'$ has mean value and variance given by

$$\langle U\rangle = \langle I\rangle\, T \tag{7.114}$$

$$\langle(\Delta U)^2\rangle = \iint_0^T |\Gamma(t-t')|^2\, \mathrm{d}t\, \mathrm{d}t'. \tag{7.115}$$

When T is very small compared with the coherence time τ_c one obtains

$$p(U) = (1/\langle U\rangle) \exp\{-U/\langle U\rangle\}, \tag{7.116}$$

or

$$p(I) = (1/\langle I\rangle) \exp\{-I/\langle I\rangle\}, \tag{7.117}$$

$$p(n) = \langle n\rangle^n/(1 + \langle n\rangle)^{1+n}, \tag{7.118}$$

and

$$\langle n\rangle = \alpha\langle I\rangle\, T, \tag{7.119}$$

$$\langle(\Delta n)^2\rangle = \langle n\rangle\, [1 + \langle n\rangle], \tag{7.120}$$

$$\langle n!/(n-m)!\rangle = m!\, \langle n\rangle^m. \tag{7.121}$$

Finally, the conditional probability $p_c(t/\tau)\,d\tau$ that a photoelectric count be registered in a time interval $d\tau$ at $t+\tau$, given that one count has been registered at time t, and assuming that the light is stationary, is given by

$$p_c(t/\tau) = \alpha\langle I(t)\, I(t+\tau)\rangle / \langle I(t)\rangle. \tag{7.122}$$

This probability equation is well suited to observing the bunching effects of photons. For polarised thermal light it becomes

$$p_c(t/\tau) = \alpha\langle I\rangle\,[1 + |\gamma(\tau)|^2]. \tag{7.123}$$

For very large T, compared with τ_c, one obtains:

$$p(U) = \delta\,(U - \langle U\rangle) \tag{7.124}$$

$$p(I) = \delta\,(I - \langle I\rangle) \tag{7.125}$$

$$p(n, T) = [\langle n\rangle^m/n!]\,\exp\,(-\langle n\rangle) \tag{7.126}$$

which is the Poisson distribution with

$$\langle(\Delta n)^2\rangle = \langle n\rangle \tag{7.127}$$

and

$$\langle n!/(n-m)!\rangle = \langle n\rangle^m. \tag{7.128}$$

For arbitrary T approximations must be used. In particular, Glauber[109] examined more carefully the problems of counting distributions for $T \geqslant \tau_c$ for the case of Gaussian noise of Lorentzian spectral profile and found that the relevant quantity, rather than being T/τ_c, is in fact the ratio of $\langle n\rangle$ and T/τ_c. If $\langle n\rangle \ll T/\tau_c$ then the counting distribution follows a Poisson law, as expressed in equations (7.126)-(7.128). If, however, the source is very intense, so that $\langle n\rangle \gg 1$ it is possible that, even though $\tau_c/T < 1$, the parameter $\mu = \langle n\rangle\,\tau_c/T > 1$.

In this case the counting distribution observed for Gaussian light deviates from a Poisson law and is given asymptotically by

$$p(n) = \langle n\rangle(2\pi n)^{-1/2} n^{-3/2}\,\exp\,[-(1/2\mu)(n^{1/2} - \langle n\rangle\, n^{-1/2})^2]. \tag{7.129}$$

7.9.2 *Laser light*

An *ideal* laser emits light of well stabilised intensity. This implies that, even i: fluctuations in the phase are possible, the amplitude of the wave field remains constant. Therefore

$$p(I) = \delta\,(I - \langle I\rangle), \tag{7.130}$$

$$p(n, T) = \langle n\rangle^n/n!, \tag{7.131}$$

$$p_c(t/\tau) = \alpha\langle I\rangle. \tag{7.132}$$

These results were all discussed during the 1960s, and several laboratories started to plan experimental verification.

Intensity fluctuations can be studied using two different techniques. In the first one the current pulses which are produced by single photoelectrons emitted by a photocathode are detected. In this kind of pulsed technique one is interested in finding the distribution of time intervals between the arrivals of successive pulses of photoelectrons and also in investigating the number of pulses registered as a function of observation time. From these data spectra and moments can be derived. This kind of experiment is usually called *photon counting* or *photoelectron counting.*

The second group of experimental methods is based on analogue techniques, using the continuous photocurrent produced by the superposition of many photoelectron pulses without trying either to resolve the discrete pulses or to measure the time intervals between pulse arrivals. These techniques are usually referred to as intensity correlation methods.

Three basic kinds of photon counting experiments have been undertaken in the investigation of intensity fluctuations. In the first type the probability distribution $p_T(n)$ that a number n of pulses is counted in a given time interval T is determined.

In the second kind of experiment, the second- and higher-order moments of the distribution are studied as a function of T.

In the third type the joint probability that n_1 photoelectrons are counted in the time interval T at t_1 and n_2 after time T at t_2 is studied.

Intensity correlation methods are in essence similar to the Hanbury Brown and Twiss experiment.

It was the development of photomultipliers and of the fast, transistorised electronic circuitry needed for data processing which had occurred in the 1950s which made the use of photon counting techniques possible.

Initial experiments, begun in 1964, demonstrated the amplitude-stabilised nature of the well-above-threshold laser output by using single detector measurements undertaken on gas lasers[123–125]. The excess noise was shown to be much smaller than that expected for a Gaussian noise source of comparable bandwidth. The effect of noise in the gas discharge on the intensity fluctuations of a gas laser was also demonstrated[123, 125, 126]; such extraneous noise masked the underlying intensity fluctuations.

Prescott and van der Ziel[126] measured excess noise on the 6328 Å line of the He-Ne laser just above threshold and observed that both the average output noise power P_n, and the coherence time of the noise τ_n decreased with increasing laser output power. More extensive measurements, both above and below threshold, were made the following year by Freed and Haus[127]. Their results are shown in figure 7.4, where the ratio S_e/S_s of excess noise to shot noise is shown.

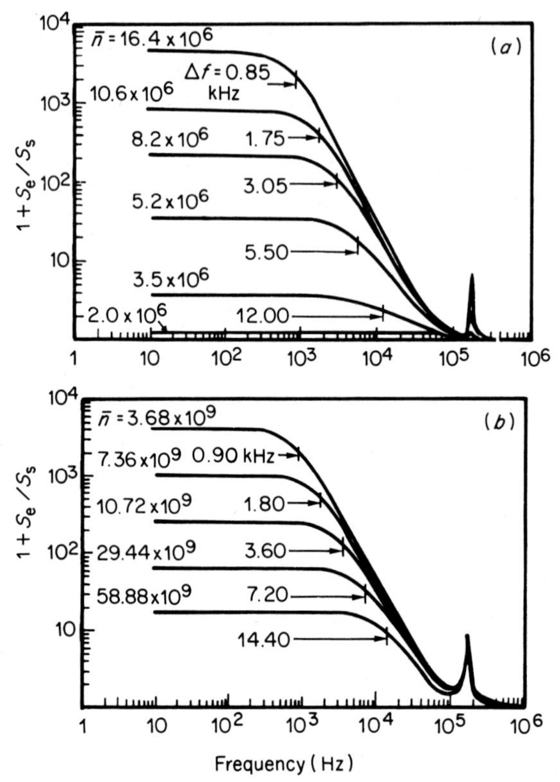

Figure 7.4 Photomultiplier current spectra observed for a gas laser. (*a*) Below threshold operation, (*b*) above threshold operation. $\bar{n}$ is the average photoelectron emission rate per second at the photocathode[127]

As threshold is approached from below (see figure 7.4(*a*)), the bandwidth of excess noise decreases from the passive cavity value of 470 kHz to 0.85 kHz for the curve closest to threshold. At the same time the peak excess noise increases. As the laser excitation was increased above threshold (figure 7.4(*b*)) the bandwidth increased and the peak noise decreased. The results above threshold were compared with the theory of noise in Van de Pol oscillators[128] and confirmed the theoretical picture of the gas laser as an oscillator perturbed by spontaneous emission[129].

Armstrong and Smith considered GaAs lasers operated continuously at 10 K[130–132]. In the first of these papers[130], received by *Physical Review Letters* on 19 November 1964, they observed intensity fluctuations by varying the injection current from a value below the threshold for coherent oscillation to a value well above threshold. The spectrum of the lasers studied had a single family of axial modes whose envelope narrowed in the manner to be expected for a homogeneously broadened fluorescence line. Only the strongest mode lased; the power in the other modes became saturated at threshold. As the injection current was varied the change in noise properties of the single-mode output was studied. Below threshold they found that the mode emitted random noise like a narrow-band black-body source; above threshold its noise was characteristic of a damped, amplitude-stabilised oscillator. Measurements of intensity fluctuations using the two-detector coincidence-counting technique of intensity interferometry developed by Hanbury Brown and Twiss were also made.

In a subsequent paper received by *Physics Letters* on 3 April 1965[132] they measured the noise in the single lasing mode using the single detector method of excess photon noise, and found excellent quantitative agreement between the two methods. Typical values of the relative fluctuations ρ for the three strongest modes derived from single detector measurements are shown in figure 7.5. Only mode 1 was lasing, and its noise goes through a peak at threshold and then falls to very low values above threshold. The noise for the two other non-lasing modes (2 and 3 in figure 7.5) remains high above threshold, as expected. The

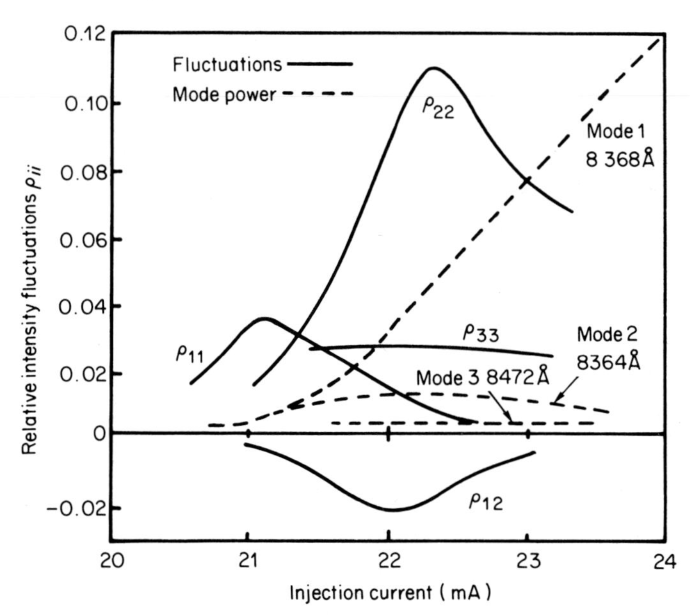

Figure 7.5 Variation with injection current of the relative intensity fluctuations ρ_{ii} for the lasing and two non-lasing modes, derived from single detector measurements. Also shown is the current dependence of the correlation ρ_{12} between the intensity fluctuations in the first (lasing) and the second (non-lasing) modes[132].

noise for mode 2 does not increase indefinitely but instead goes through a maximum, because of coupling between modes. This coupling also implies a correlation between noise in different modes, which is verified by the non-zero value of ρ_{12} in figure 7.5.

The last paper[133] gave a full account of all these measurements.

Later on, in 1966, Geusic[134] investigated intensity fluctuations in a YAl garnet : Nd laser operating cw near threshold.

The problem of experimentally determining the form of the distribution law of photons was now exposed, and ready to be tackled. The first work, still not at the experimental stage, but clearly pointing towards an experiment, was by McLean and Pike[135] and was received by *Physics Letters* on 19 March 1965. In this paper the authors gave an explicit expression for the counting distribution $p(n, t)$ for Gaussian light in the region which is of practical interest, when $T \gg \tau_c$; and this under conditions which can normally be obtained in practice.

At nearly the same time in France efforts were being made to measure some of the statistical properties of laser light. In a short note J Marguin *et al*[136] presented some experimental evidence that the ratio $\langle \Delta n^2 \rangle / \langle n \rangle$ was around unity when the counting time T was shorter than the inverse bandwidth of a He–Ne laser. This result, however, attracted little attention and was not pursued.

In June 1965 there was a Conference in San Juan, Puerto Rico, on *The Physics of Quantum Electronics.* At this conference important papers on photon-counting were presented: a theoretical paper by Glauber[109]; and three experimental papers by Johnson, McLean and Pike[137], by Freed and Haus[138] and by Armstrong and Smith[131]. In these last papers photon counting experiments are described. H A Haus[139] had already presented his results in a report to MIT the year before, and E R Pike had announced the experiment under way at Malvern in a research report bulletin early in March 1965[140].

Freed and Haus[138] considered the normalised factorial moments

$$F(k) = [\overline{N(N-1)\ldots(N-k+1)} - \overline{N}^k]/(\overline{N})^{k-1}, \qquad (7.133)$$

below and above threshold of a He–Ne gas laser as a function of counting interval, up to $F(4)$.

The assumption made by the authors was that below threshold the amplitude of the electric field components of the laser output had a Gaussian probability distribution of zero mean, and that above threshold, operation of the laser occurred in one single mode. With these assumptions they calculated the expected $F(k)$ and then made a comparison with experiment (figure 7.6), which gave rather good agreement between experiment and theory.

Pike[137] reported . . .

> In essence, we measure the number of photons counted in a time T (using the photomultiplier for single-electron counting rather than as a

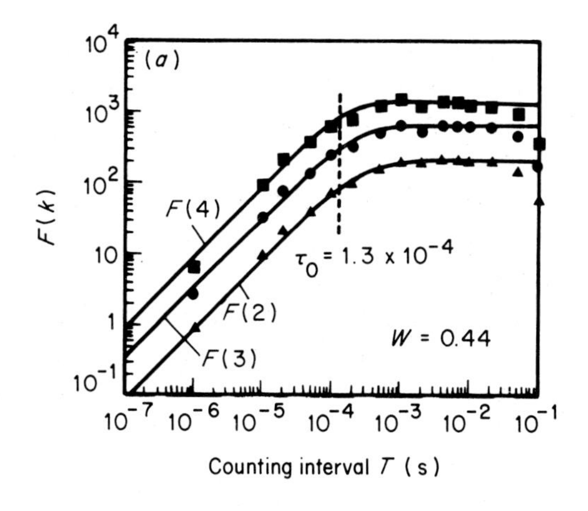

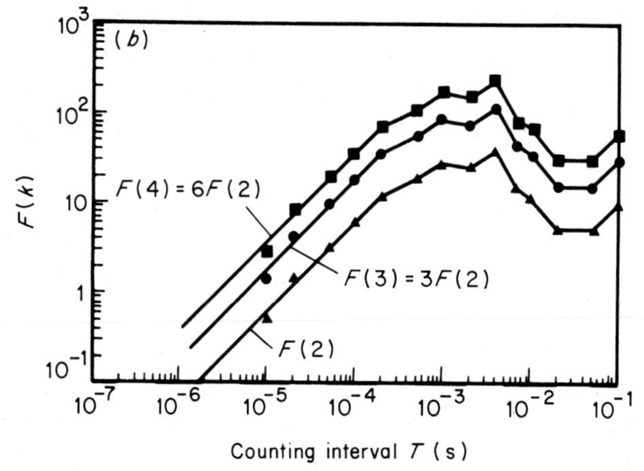

Figure 7.6 Normalised factorial moments: (*a*) below threshold (photon rate $2.81 \times 10^9 \text{s}^{-1}$; photoelectron rate $4.31 \times 10^6 \text{s}^{-1}$), (*b*) above threshold (photon rate $8.15 \times 10^{11} \text{s}^{-1}$; photoelectron rate $11.09 \times 10^6 \text{s}^{-1}$)[138].

current multiplier), store this number away electronically, and repeat a large number of times.

It is easily shown that to measure a shift of variance of one part in 10^3 of a Poisson distribution, with 95% confidence, one needs to sample the distribution to the order of 10^7 times.

The variances of photon counting distributions have been used in two recent investigations, but only small numbers of samples were obtained. In our experiments the number of samples obtained has ranged from 5×10^7 to over 10^{10}, taking times from a few minutes to several hours.

> In a number of circumstances one can calculate explicitly not only the excess photon noise to be expected but the actual photon-counting distribution. With the numbers of samples which we accumulate, this distribution can be found accurately, and as we shall show, the form of such distributions when compared with the expected results leads to a great deal of experimental information which would be quite masked in a simple extraction of the variance or a noise measurement.

For the experimental analysis they worked in terms of the quantity

$$F(n) = [(n+1)\,p(n+1, T)]/[p(n, T)] = \langle n \rangle (1 - \alpha\delta) + \alpha\delta\, n, \quad (7.134)$$

where α was the quantum efficiency and δ the degeneracy parameter.

In these first photon-counting distribution measurements presented at Puerto Rico a number of new technical considerations prevented them obtaining exactly the expected results. Among these were the necessarily finite resolution time; the source and circuit stability; and multiplier afterpulsing. After mastering these problems distributions were obtained which permitted moments as high as n^6 to be later successfully compared with theory for thermal and laser sources[141].

In November 1965 two papers were submitted to *Physics Review Letters* almost at the same time. On 18 November a paper by Arecchi[142] arrived in which the experimental distribution $p(n, T)$ was given for an artificial Gaussian source and for a laser, together with the measurement of the first three moments of the photon counting distribution. Surprisingly good agreement, considering the operating conditions, was found with the theoretical predictions.

The following day (19 November), Freed and Haus[143] presented the photon statistics for a laser below threshold (that is, a narrow-band Gaussian source).

The experiment described by Arecchi and performed by a group of researchers at CISE, Milan, was done by sending the light of an amplitude-stabilised single-mode He–Ne laser onto a moving, ground-glass disc. The effect of diffraction by the many tiny moving scattering centres was to convert the scattered beam into Gaussian noise whose bandwidth is determined by the velocity of the scattering plate.

A similar procedure was first introduced by Martienssen and Spiller[144]. Photo-electron distributions for this synthetised Gaussian source and for the laser source were given.

Freed and Haus in the paper[143] presented the first measurement of laser light below threshold. This work was stimulated by the determination of the probability distribution of photoelectron counts reported in Puerto Rico by Johnson *et al*[137]. Freed and Haus said:

> A determination of the probability distributions of photoelectron counts has been reported by Johnson, McLean and Pike using an

incoherent light source. Their work provided the initial impetus for the measurements reported here.

The laser, below threshold, was considered to produce a narrow-band Gaussian light. Figure 7.7 shows the experimental results. If the laser output were to consist of a pure narrow-band Gaussian light centred at a single

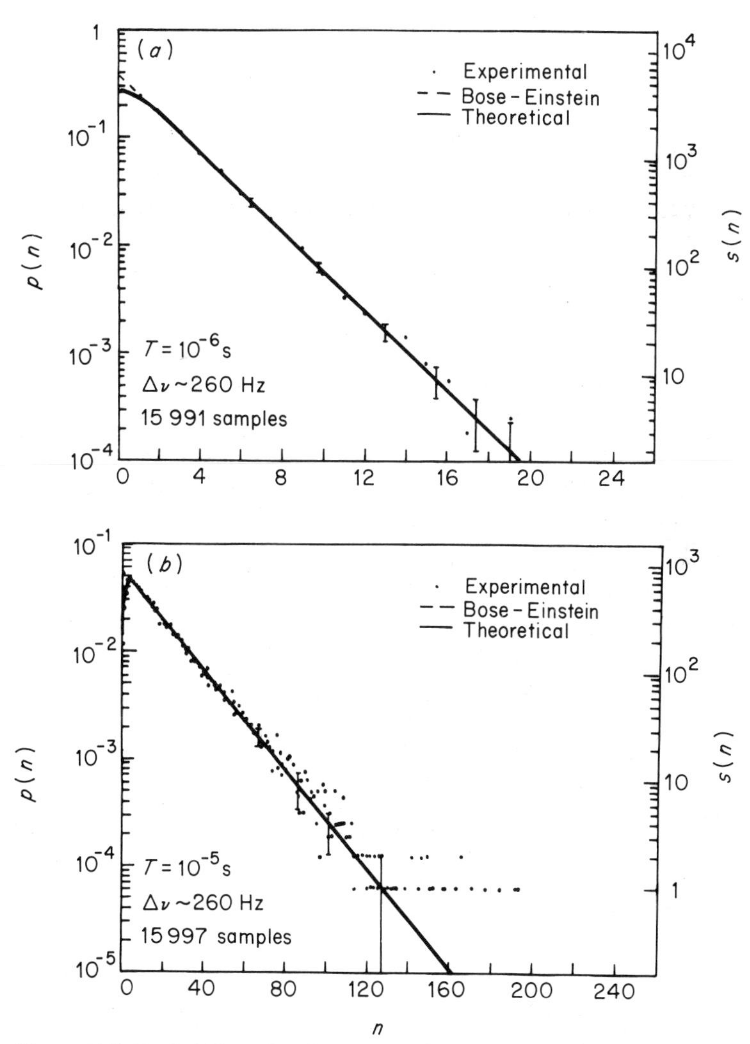

Figure 7.7 Probability distribution and number of observed samples versus photoelectron count for a laser below threshold: (*a*) counting interval $T < \tau_c$[148], (*b*) $T > \tau_c$[143]; copyright © 1966 IEEE.

frequency, the probability of observing n photoelectrons within a time interval which was short compared with the inverse bandwidth, would follow the Bose-Einstein distribution law[118] and, on the semilog plot of figure 7.7(a) this should appear as a straight line. One can see from the figure that the experimental points lie on or about a straight line down to counts of the order of four. By assuming that the deviation is due to admixture of other modes at other frequencies, Freed and Haus were able to calculate the full line also shown in figure 7.7(a).

The spectral width of the source was such that it was easy to satisfy the requirement $T \ll \tau_c$, as shown in figure 7.7(a). Figure 7.7(b) is, instead, undertaken with a counting interval longer than the coherence time. Under this condition it was expected that the asymptotic expression (7.129), developed by Glauber[109] would fit, as actually happened.

A more exhaustive account of the CISE measurements was published later[145], together with a few other works[146, 147], until April 1966, when photon counting experiments were debated at the *4th International Quantum Electronics Conference, Phoenix, Arizona.* Papers by Freed and Haus[148], and Arecchi[149] were submitted which, basically, gave a general review of the results previously obtained. Inexplicably, the Conference Committee did not accept the contribution from the Malvern group.

Meanwhile, the theoreticians were at work. Nearly all the theories of laser noise up to 1964 were based on a linear approach[150]. This approach predicts that the statistical properties of laser light are the same as those of light from thermal sources. The amplitude should have a Gaussian distribution. In 1964 a non-linear approach was developed almost simultaneously by Haken[151] and Lamb[152]. A non-linear laser equation which was able to predict the field fluctuations was thus derived, and it was demonstrated that laser light below threshold is narrow-band Gaussian, and that above threshold the laser acts as a self-sustained oscillator with a high stabilised (classical) amplitude.

These results were confirmed by Armstrong and Smith[130, 153], and Freed and Haus[127].

The change in photon statistics between sub-threshold and above threshold operation was quantitatively predicted by Risken in 1965[154], and then confirmed by the experiments of Armstrong and Smith[155], and others[156].

Later, in 1966, Lamb *et al* finally established the quantum theory of lasers[157].

Light from a laser operating somewhat above threshold does exhibit intensity fluctuations and in this region the laser field may be approximated by a linear superposition of an amplitude-stabilised wave and a Gaussian noise wave. The case was first treated by Lachs[158] and Glauber[109] using the quantum theory, and by Mandel and Wolf[159] classically. An approach based on the calculation of the correlation functions was also developed by Morawitz[160] and generalised by Peřina[161].

The accurate experimental verification of the photon counting distribution was hindered by many practical difficulties. Most of them were solved in subsequent years and the work of the RRE group of Malvern was fundamental in this respect[162]. The group had, in fact, been working on the problem from the beginning, as we have seen; the importance of their work was quickly recognised by Kastler who, on seeing one of their experimental distributions displayed, commented with delight, 'Ah, la loi des Bosons!' and he reproduced what is probably the first photon counting distribution to be published in a text-book in these very early days[163].

Dead-time effects, which had strongly affected the first measurements of $F(n)$, equation (7.134)[137], were corrected with a method discussed in Johnson *et al*[147], and were published in a paper received on 31 May 1966 by *Optica Acta*[164]. Figure 7.8 shows the large, negative slopes of $F(n)$ versus n curves as obtained by Johnson *et al*[137] for a stabilised single-mode gas laser (circles) and a stabilised tungsten lamp (triangles), and the dead-time modified Poisson distributions (broken lines) which fit the experimental data excellently.

Other groups were also working on the problem of radiation fluctuations and photon statistics. In Rochester, Professor H Gamo and R E Grace were studying intensity fluctuations in single-mode gas lasers near the oscillation threshold with an ingenious two-photomultiplier technique. Their results were first presented in the *Second Rochester Conference on Coherence and Quantum Optics* 24 June 1966[165].

Chang *et al* in 1967[166] measured fluctuation statistics up to the fourth order, and found good agreement with the non-linear oscillator theory.

Using a three-detector coincidence-counting technique, Davidson and Mandel[167] observed time correlation functions of amplitude fluctuations up to the sixth order and confirmed earlier observation of fourth-order field statistics[168]. Later Meltzer, Davis and Mandel[169] observed the counting distribution $p_T(n)$ for a range of values of the pump parameter encompassing the threshold region, and for a number of different ratios of counting time to correlation time.

Jakeman *et al*[170] measured moments up to fourth order over a similar range of parameters.

In 1966 photon bunching in time in a thermal light beam was also demonstrated. This bunching in time of a photon beam is completely analogous to the well known tendency of a photon gas or indeed any boson gas to form clusters in space[171]. The bunching phenomena in a photon beam was also analysed by Dicke in 1954[172] in terms of the radiation process in an excited gas. He showed that there is a correlation between the emission direction of successive photons,

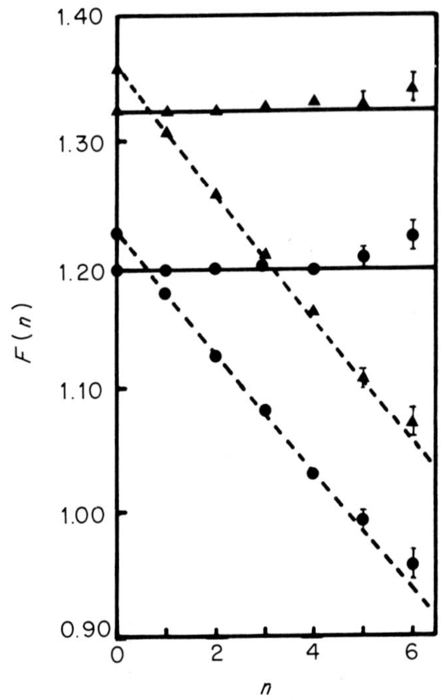

Figure 7.8 The experimentally measured and dead-time-corrected values of $F(n)$ for a stabilised single-mode gas laser (circles) and a stabilised tungsten lamp (triangles). The data in each case were based on $\sim 10^8$ samples taken with a 1 μs sampling l time. The broken lines are a fit of dead-time-modified Poisson distributions to the observed data. The horizontal sets of points are the experimental data corrected for dead-time effects and the associated full horizontal lines are for Poisson distributions of the same mean. The calculated dead-time in each case is 19.38 ns[164].

such that the emission probability in the same direction is nearly twice that in an arbitrary direction.

The first measurement of bunching of photons from a thermal source was done in 1966 by Morgan and Mandel[173] who used a low-pressure ^{198}Hg gas discharge lamp. Light was passed through a pinhole (diameter 0.54 mm), an optical filter that isolated the blue 5461 Å line, and a linear polariser; it finally fell on a 56 AVP photomultiplier through a rectangular aperture (0.37×0.47 mm^2) whose dimensions were small enough to ensure a degree of coherence of at least 90% across the beam. The pulses from the photomultiplier were processed in such a way as to enable measurement of the conditional probability $p(t/\tau)$ as τ was varied between 1 and 6 ns.

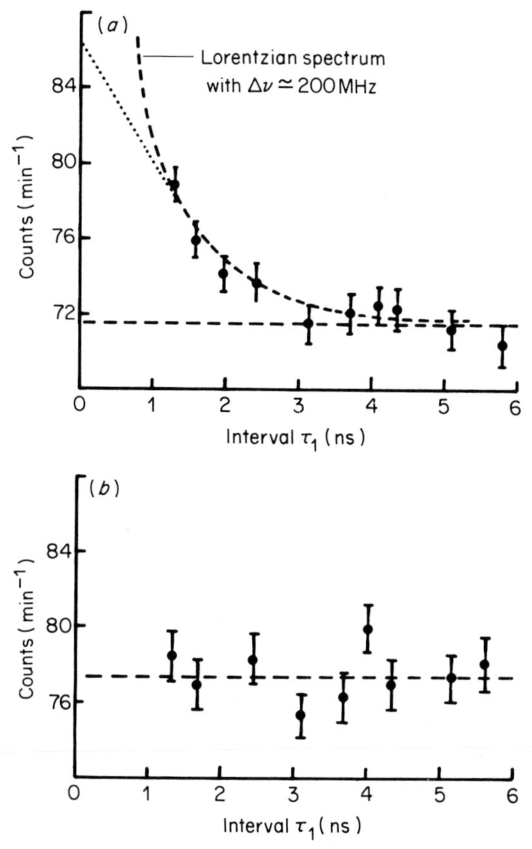

Figure 7.9 Counting rates illustrating the phenomenon of photon bunching. In (*a*) with light from a ^{198}Hg source; in (*b*) with light from a tungsten lamp. The ordinates represent essentially a quantity which is proportional to the integral $\int_{\tau_1}^{\tau_2} p_c(\tau)$ where τ_2 is a constant[173].

Results are shown in figure 7.9(*a*). For comparison the results of similar measurements carried out with a tungsten lamp as source are shown in figure 7.9(*b*). The results in this case showed no significant bunching of photon counts since the coherence time was far shorter than 1 ns.

Photon bunching is an aspect of the Bose–Einstein statistics, i.e. of the statistics followed by photons emitted in equilibrium conditions. It is absent in light emitted by a coherent source (as in a laser[145]) and it is also absent in chaotic sources when fluctuations are smeared out because of counting in time intervals larger than the coherence time.

In this respect we observe that the Poisson distribution is to be expected when counting well stabilised laser light and when counting the light from a chaotic source with $T \gg \tau_c$. This may at first sight appear somewhat confusing. In the case of the laser a Poisson distribution corresponds to the genuine absence of intensity fluctuations; in the case of a chaotic source the Poisson distribution occurs because in the measuring time $T(\gg \tau_c)$ the detector averages out all the fluctuations which are in fact present in this case, and which appear as soon as $T \leqslant \tau_c$ and give rise to photon bunching, etc[174].

When photons are emitted in non-equilibrium conditions other behaviour may occur: an example is antibunching, which was first observed by Mandel and co-workers[175].

The importance of the statistical properties of radiation is now widely appreciated in many different applications, for example, photon-correlation methods in scattering, propagation in the turbulent atmosphere, non-linear optics, etc[176].

Notes

1. E Whittaker, *A History of the Theories of Aether and Electricity* (Harper and Brothers: New York, 1960)
2. M Jammer, *The Conceptual Development of Quantum Mechanics* (McGraw-Hill: New York, 1966)
3. M Planck gave his formula in a communication read on 19 October 1900 before the German Physical Society and published in *Verh. Deutsch. Phys. Ges.* **2** (1900) p202
4. Lord Rayleigh, *Phil. Mag.* **49**, 539 (1900); cf *Nature* **72**, 54 (May 1905); *Nature* **72**, 243 (July 1905); J H Jeans, *Phil. Mag.* **10**, 91 (1905)
5. This result was read by Planck on 14 December 1900 before the German Physical Society and published in *Verh. Deutsch. Phys. Ges.* **2** (1900) p 237. This and the paper in note 3 were re-edited and printed in the new form in *Ann. Phys., Lpz.* **4** (1901) p 553. An excellent commentary to Planck's papers was given by M J Klein in vol 1 of the *Archive for History of Exact Sciences*, (1962) p459
6. A Einstein, *Ann. Phys., Lpz.* **17**, 132 (1905). English translations are by A B Arons and M B Peppard: *Am. J. Phys.* **33**, 367 (1965) and D ter Haar, *The Old Quantum Theory* (Pergamon: Oxford, New York, 1967) p 91. It was nominally for this work that the Nobel Prize was awarded to Einstein.
7. The three papers are: A Einstein, *Über einen die Erzeugung und Verwandlung des Lichtes betreffenden heuristischen Gesichtspunkt* of which in note 6 here; A Einstein, *Die von der molekularkinetischen Theorie der Wärme geforderte Bewegung von in ruhenden Flüssigkeiten suspendierten Teilchen, Ann. Phys., Lpz.* **17**, 549 (1905). This paper was reprinted in A Einstein, *Investigations on the Theory of the Brownian Movement*, translated by A D Cowper, with notes by R Fürth (Dover: New York, 1956); A Einstein *Zur Elektrodynamik bewegter Körper, Ann. Phys.,*

Lpz. **17**, 891 (1905). This paper was reprinted in A Einstein, H A Lorentz, H Weyl and H Minkowski, *The Principle of Relativity: A Collection of Original Memories on the Special and General Theory of Relativity* translated by W Perrett and G B Jeffrey, with notes by A Sommerfeld, (Dover: New York)

8 A deeper analysis of this paper is in Martin J Klein *Einstein's First Paper on Quanta* in the *Natural Philosopher* (Baisdell: New York, 1963) vol 2 p59 and in *Einstein and the Wave-Particle Duality*, ibid., vol 3 (1964) p1

9 Planck proposed (*Verh. Deutsch. Phys. Ges.* **13**, 138 (1911)) a new hypothesis that, although emission of radiation always takes place discontinuously in quanta, absorption on the other hand is a continuous process which takes place according to the laws of the classical theory. Radiation while in transit might therefore be represented by Maxwell's theory, and the energy of an oscillator at any instant might have any value whatever. For a detailed discussion of these important steps the reader is referred to note 2

10 G N Lewis, *Nature* **118**, 874 (1926)

11 P Lenard, *Ann. Phys., Lpz.* **8**, 149 (1902); see also J H Jeans, *Report on Radiation and the Quantum Theory* (The Electrician Printing and Publishing Co: London, 1914); R H Stuewer in *Historical and Philosophical Perspectives of Science*, edited by R H Stuewer (University of Minnesota Press: Minneapolis, 1970) p246

12 O W Richardson and K T Compton, *Phil. Mag.* **24**, 575 (1912)

13 A L Hughes, *Phil. Trans.* **212**, 205 (1912)

14 R A Millikan, *Phys. Rev.* **7**, 355 (1916); see also R A Millikan and C F Eyring, *Phys. Rev.* **27**, 51 (1926); R A Millikan and C L Lauritsen, *Proc. Natl Acad. Sci. USA* **14**, 45 (1928); R A Millikan, *Phys. Rev.* **18**, 236 (1921)

15 C A Sadler and P Mesham, *Phil. Mag.* **24**, 138 (1912)

16 A H Compton, *Phil. Nat. Res. Council USA* vol 4 no 20 October 1922; see also *Phys. Rev.* **21**, 483 (1923)

17 cf also P Debye, *Phys. Z.* **24**, 161 (1923) who discovered the theory independently; A H Compton, *Phil. Mag.* **46**, 897 (1923); A H Compton and W A Simon, *Phys. Rev.* **25**, 306 (1925)

18 A Einstein, *Phys. Z.* **10**, 185 (1909)

19 H A Lorentz, *Les Theories Statistiques en Thermodynamique* (G B Teubner: Leipzig, 1916) p114

20 A Einstein, *Phys. Z.* **18**, 121 (1917)

21 For a brief general survey of these papers see for example W Pauli in *Albert Einstein: Philosopher-Scientist* edited by P A Schilpp (Tudor: New York, 1957) p147

22 P Ehrenfest, *Ann. Phys., Lpz.* **36**, 91 (1911)

23 A Joffé *Ann. Phys., Lpz.* **26**, 534 (1911)

24 L Natanson, *Phys. Z.* **12**, 659 (1911)

25 G Krutkow, *Phys. Z.* **15**, 133 (1914)

26 L de Broglie, *Comptes Rendus* **175**, 811 (1922)

27 L de Broglie, *J. Physique* **3**, 422 (1922)

28 W Bothe, *Z. Phys.* **20**, 145 (1923)

29 More recently J H Webb, *Am. J. Phys.* **40**, 850 (1972) has observed that if stimulated emission is omitted in the Einstein derivation, the Wien formula results, and the fluctuations become equal to the single term n, which is characteristic of classical statistics. This shows that stimulated emission is the mechanism responsible for the appearance of the Bose–Einstein statistics in black-body radiation

30 S N Bose, *Z. Phys.* **26**, 178 (1924), English translation in *Am. J. Phys.* **44**, 1056 (1976)

31 See W A Blanpied, *Am. J. Phys.* **40**, 1212 (1972); M Jammer note 2, p248

32 Bose was able to derive Planck's formula using only the light quanta hypothesis combined with statistical mechanics. He therefore succeeded in a task Einstein had failed in for 15 years, overcoming the logical defect existing in previous derivations where quantum concepts were used together with classical electrodynamics which was already demonstrated to be inconsistent with them (see Chapter 2)

33 A Einstein, *Berlin Sitz.* p 261 (1924)

34 S N Bose, *Z. Phys.* **27**, 384 (1924), English translation by O Theimer and B Ram in *Am. J. Phys.* **45**, 242 (1977)

35 This new paper by Bose treated the equilibrium of the radiation when it is in interaction with matter and is continuously emitted and absorbed. In Chapter 2 we have seen Einstein's derivation of Planck's law (here note 20) which assumes that the rate of emission contains a term proportional to the photon number in one phase cell for the rate of spontaneous transition (stimulated emission). This special model was extended to multilevel systems by Einstein and Ehrenfest (cf Chapter 2). On the other hand Pauli (*Z. Phys.* **18**, 272 (1923)) had studied light scattering from electrons and had shown that if the Compton effect was taken into account, Planck's distribution was stationary; Bose in his paper (note 34) considered the general case and showed that in his formulation both Pauli processes and Einstein–Ehrenfest processes were included as special cases, and this alone was already a noteworthy result. Further Bose observed that instead of Einstein's hypothesis of spontaneous and stimulated decays, for emission it is possible to consider only spontaneous emission if the absorption is taken to be proportional, not to the number of quanta per phase space cell, but to this number divided by this same number plus one. Until one considers radiative equilibrium, this hypothesis is as good as Einstein's because only these ratios appear

36 A Einstein, letters to P Ehrenfest dated 29 September, 2 December 1924, Einstein Archive

37 M J Klein, *Natural Phil.* **3**, 26 (1963)

38 A Einstein, *Berlin Sitz.* pp 3, 18 (1925)

39 For a discussion of papers 33 and 38 see M J Klein in *The Natural Philosopher* vol 3 (Blaisdell: New York, 1964) p 1

40 A Smekal, *Z. Phys.* **37**, 319 (1926)

41 W Bothe, *Z. Phys.* **41**, 345 (1927)

42 R Furth, *Z. Phys.* **48**, 323 (1928)

43 R Furth, *Z. Phys.* **50**, 310 (1928)

44 R E Burgess, *Proc. Phys. Soc.* **53**, 293 (1941)

45 R E Burgess, *Proc. Phys. Soc.* **58**, 313 (1946)
46 J M W Milatz and H A van der Velden, *Physica* **10**, 369 (1943)
47 W B Lewis, *Proc. Phys. Soc.* **59**, 34 (1947)
48 R C Jones, *J. Opt. Soc. Am.* **37**, 879 (1947)
49 P Fellgett, *J. Opt. Soc. Am.* **39**, 970 (1949)
50 H B Callen and T A Welton, *Phys. Rev.* **83**, 34 (1951)
51 H Ekstein and N Rostoker, *Phys. Rev.* **100**, 1023 (1955)
52 J Weber, *Phys. Rev.* **101**, 1620 (1956)
53 H Nyquist, *Phys. Rev.* **32**, 110 (1928); J B Johnson, *Phys. Rev.* **32**, 97 (1928)
54 R C Jones, *Adv. Electron.* **5**, 1 (1953)
55 P Fellgett, *Nature* **179**, 956 (1957)
56 R Q Twiss and R Hanbury Brown, *Nature* **179**, 1128 (1957)
57 P Fellgett, R C Jones and R Q Twiss, *Nature* **184**, 967 (1959)
58 D K C MacDonald, *Rep. Prog. Phys.* **12**, 56 (1948)
59 E Verdet, *Ann. Sci. l'Ecole Normale Supérieure* **2**, 291 (1865)
60 L Mandel and E Wolf, *Rev. Mod. Phys.* **37**, 231 (1965)
61 M Born and E Wolf, *Principles of Optics* (Pergamon: Oxford, 1965) ch 10
62 The concept of an analytic signal was introduced by D Gabor, *J. Instrum. Electron. Engrs.* **93**, part 4, 429 (1946). See also V I Bunimovich, *J. Tech. Phys. USSR* **19**, 1231 (1949). It is well treated in the classical textbook by M Born and E Wolf (note 61)
63 F Zernike, *Physica* **5**, 785 (1938). Measure of correlation of light vibrations was introduced by M von Laue, *Ann. Phys., Lpz.* **23**, 1, 795 (1907). Joint probability distribution for the light disturbances was also determined by P H van Cittert, *Physica* **1**, 201 (1934); ibid. **6**, 1129 (1939). See also L Janossy, *Nuovo Cim.* **6**, 111 (1957), ibid. **12**, 369 (1959). Actually Zernike introduced what is now called the mutual intensity function, that is Γ_{12} $(\tau = 0)$. The function $\Gamma_{12}(\tau)$ was first introduced by E Wolf, *Proc. R. Soc.* A **230**, 246 (1955)
64 H H Hopkins, *Proc. R. Soc.* A **208**, 263 (1951); ibid. A **217**, 408 (1953)
65 E Wolf, *Proc. R. Soc.* A **230**, 246 (1955); ibid. A **225**, 96 (1954); *Nuovo Cim.* **12**, 884 (1954). Here the definitions in term of analytic signal are introduced
66 A Blanc-Lapierre and P Dumontet, *Rev. d'Optique* **34**, 1 (1955)
67 The publication *Photoelectric Mixing of Incoherent Light* was in *Phys. Rev.* **99**, 1691 (1955) under the names A T Forrester, R A Gudmundsen and P O Johnson. Early suggestions were A T Forrester, W E Parkins and E Gerjnoy *Phys. Rev.* **72**, 728 (1947); E Gerjnoy, A T Forrester and W E Parkins, *Phys. Rev.* **73**, 922 (1948)
68 C T J Alkemade, *Physica* **25**, 1145 (1959)
69 A T Forrester, *J. Opt. Soc. Am.* **51**, 253 (1961); see also *Advances in Quantum Electronics* edited by J R Singer (Columbia University Press: New York, 1961) p233
70 These two approaches are sometimes referred to in the literature under the names of *heterodyne* and *homodyne* detection. This terminology is, however, incorrect. The term *heterodyne* should be used when beats are produced between two different lasers. When, as is usually the case in a scattering experiment, beats are produced between light from the same

laser which is split in a *scattered* beam and a *reference* beam, the term *homodyne* should be used, while when the same beam beats with itself one should speak of *self-beating* spectroscopy

71 A Javan, W R Bennett and D R Herriott, *Phys. Rev. Lett.* **6**, 106 (1961) applied the technique to study He–Ne laser output; later A Javan, E W Ballik and W L Bond, *J. Opt. Soc. Am.* **52**, 96 (1962) used the beat note produced by the output of two independent lasers which simultaneously illuminated a single photomultiplier to determine the stability of the lasers. In the same year S E Harris, B J McMurtry and A E Siegman, *Appl. Phys. Lett.* **1**, 37 (1962) used beats in solid-state devices and applied the method to study ruby-laser beats, B J McMurtry and A E Siegman, *Appl. Opt.* **1**, 51 (1962)

72 See for example H Z Cummins and H L Swinney, in *Prog. Opt.* edited by E Wolf, **8**, 133 (1970); E Jakeman, C J Oliver and E R Pike, *Adv. Phys.* **24**, 349 (1975)

73 R Hanbury Brown, R C Jennison and M K Das Gupta, *Nature* **170**, 1061 (1952)

74 R Hanbury Brown and R Q Twiss, *Phil. Mag.* Ser 7 **45**, 663 (1954)

75 R C Jennison and M K Das Gupta, *Phil. Mag.* Ser 8 **1**, 55 (1956)

76 R Hanbury Brown and R Q Twiss, *Nature* **177**, 27 (1956)

77 R Hanbury Brown and R Q Twiss, *Nature* **178**, 1046 (1956)

78 R Hanbury Brown and R Q Twiss, *Proc. R. Soc.* A **242**, 300 (1957)

79 R Hanbury Brown and R Q Twiss, *Proc. R. Soc.* A **243**, 291 (1957)

80 R Hanbury Brown and R Q Twiss, *Proc. R. Soc.* A **248**, 199 (1958)

81 R Hanbury Brown and R Q Twiss, *Proc. R. Soc.* A **248**, 222 (1958)

82 R Hanbury Brown, *The Intensity Interferometer* (Taylor and Francis: London, 1974)

83 Note 82, p3

84 Note 82, p4

85 Note 82, pp 6 ff

86 A Ádám, L Janossy and P Varga, *Acta Phys. Hung.* **4**, 301 (1955)

87 R Q Twiss, A G Little and R Hanbury Brown, *Nature* **180**, 324 (1957)

88 E Brannen and H I S Ferguson, *Nature* **178**, 481 (1956)

89 This result comes from the Gaussian nature of the thermal light. That white light of thermal origin has the properties of a Gaussian random process had been well established many years ago in several classical papers, see A Einstein, *Ann. Phys., Lpz.* **47**, 879 (1915). It was later shown by several authors; cf I S Reed, *IRE Trans. on Inform. Theory* **IT-8**, 194 (1962); C L Mehta, *Lectures on Theoretical Physics* edited by W E Brittin (University of Colorado Press: Boulder, Colorado, 1961) vol 7 p 398; L Mandel and E Wolf, *Phys. Rev.* **124**, 1694 (1961)

90 The relation was later derived by L Mandel, *J. Opt. Soc. Am.* **51**, 1342 (1961)

91 D Gabor, *Phil. Mag.* Ser 7, **41**, 1161 (1950); *Progress in Optics* vol 1 (North-Holland: Amsterdam, 1961) p 111; L Mandel, *J. Opt. Soc. Am.* **52**, 1407 (1962) and note 99

92 G A Rebka and R V Pound, *Nature* **180**, 1035 (1957)

93 E Brannen, H I S Ferguson, W Wehlan, *Can. J. Phys.* **36**, 871 (1958); see also W Martienssen and E Spiller, *Am. J. Phys.* **32**, 919 (1964);

G L Farkas, L Janossy, Z Náray and P Varga, *Acta Phys. Acad. Sci. Hung.* **18**, 199 (1965)

94 E M Purcell, *Nature* **178**, 1449 (1956); sentences reproduced by permission copyright © 1956 Macmillan Journals Limited)

95 See note 76; L Mandel, *Proc. Phys. Soc.* **71**, 1037 (1958); **74**, 233 (1959); F D Kahn, *Opt. Acta* **5**, 93 (1958), U Fano, *Am. J. Phys.* **29**, 539 (1961)

96 L Mandel, *Proc. Phys. Soc.* **72**, 1037 (1958)

97 L Mandel, *Proc. Phys. Soc.* **74**, 233 (1959)

98 S O Rice, *Bell Syst. Tech. J.* **23**, 282 (1944), ibid. **24**, 46 (1945)

99 L Mandel, *J. Opt. Soc. Am.* **51**, 797 (1961)

100 A Einstein, *Congres Solvay* (1912)

101 L Mandel, E C G Sudarhsan and E Wolf, *Proc. Phys. Soc.* **84**, 435 (1964)

102 R J Glauber, *Phys. Rev. Lett.* **10**, 84 (1963); see also R Hanbury Brown and R Q Twiss, *Nature* **177**, 27 (1956); G A Rebka and R V Pound, *Nature* **180**, 1035 (1957); L Mandel and E Wolf, *Phys. Rev.* **124**, 1696 (1961)

103 R J Glauber, *Phys. Rev.* **130**, 2529 (1963)

104 R J Glauber, *Phys. Rev.* **131**, 2766 (1963)

105 The density operator was introduced by von Neumann, *Gesellschaft der Wissenschaften zu Göttinger Math. Phys. Nachrichten* 245 (1927) and *Mathematical Foundations of Quantum Mechanics* (Princeton University Press; Princeton NJ, 1955); see also P A M Dirac *Proc. Camb. Phil. Soc.* **25**, 62 (1929); **26**, 376 (1930); **27**, 240 (1930). General reviews were given by U Fano, *Rev. Mod. Phys.* **29**, 74 (1957); D ter Haar, *Rep. Prog. Phys.* **24**, 304 (1961); see also W H Louisell, *Radiation and Noise in Quantum Electronics* (McGraw-Hill: New York, 1964)

106 The *P* representation had been already introduced almost contemporarily by E C G Sudarshan, *Phys. Rev. Lett.* **10**, 277 (1963); see also *Proc. Symposium on Optical Masers* (Wiley: New York, 1963) p45; J R Klauder, J McKenna and D G Currie, *J. Math. Phys.* **6**, 733 (1965); C L Mehta and E C G Sudarshan, *Phys. Rev.* **138**, B274 (1965). See also J R Klauder and E C G Sudarshan, *Fundamentals of Quantum Optics* (Benjamin: New York, 1968)

107 Coherent states are just displaced versions of the ground state of the harmonic oscillator (see note 104) and were first discussed by E Schrödinger, *Naturwiss.* **14**, 644 (1926)

108 Functions of the type of the *P* functions were first discussed by E Wigner, *Phys. Rev.* **40**, 749 (1932). The name sometimes used of *quasi probability*, was suggested by G A Backer, *Phys. Rev.* **109**, 2198 (1958)

109 R J Glauber, *Physics of Quantum Electronics (Conf. Proc., San Juan, Puerto Rico, 28–30 June 1965)* edited by P L Kelley, B Lax and P E Tannenwald (McGraw-Hill: New York, 1966) p788

110 R J Glauber, in *Quantum Optics and Electronics, Les Houches 1964* edited by C DeWitt, A Blandin and C Cohen-Tannoudji (Gordon and Breach: New York, 1964) p65

111 F Ghielmetti, *Phys. Lett.* **12**, 210 (1964)

112 L Mandel, *Phys. Rev.* **136**, 647 (1964)

113 L Mandel, *Phys. Rev.* **152**, 438 (1966)

114 L Mandel and E Wolf, *Phys. Rev. Lett.* **10**, 276 (1963)

115 E Wolf, in *Quantum Electronics* edited by P Grivet and N Bloembergen (Dunod: Paris, 1964) p13
116 L Mandel, in *Quantum Electronics* edited by P Grivet and N Bloembergen (Dunod: Paris, 1964) p101
117 K E Cahill, *Phys. Rev.* **138**, B1566 (1965)
118 U M Titulaer and R J Glauber, *Phys. Rev.* **140** B676 (1965)
119 M Bertolotti, B Crosignani and P Di Porto, *J. Phys. A: Gen. Phys.* **3** L37 (1970)
120 M Bertolotti, B Crosignani, P Di Porto and D Sette, *Phys. Rev.* **157**, 146 1967); see also Y R Shen, *Phys. Rev.* **155**, 921 (1967)
121 For a full discussion see B Crosignani, P Di Porto and M Bertolotti, *Statistical Properties of Scattered Light* (Academic: New York, 1975)
122 P Di Porto, M Bertolotti and B Crosignani, *J. Appl. Phys.* **40**, 5083 (1969); see also note 121 for a deeper discussion; P J Bourke, J Butterworth, L E Drain, P A Egelstaff, P Hutchinson, B Moss, P Schofield, A J Hughes, J J B O'Shaughnessy, E R Pike, E Jakeman and D A Jackson, *Phys. Lett.* **28A**, 692 (1969)
123 J A Bellisio, C Freed and H A Haus, *Appl. Phys. Lett.* **4**, 5 (1964)
124 R L Bailey and J H Sanders, *Phys. Lett.* **10**, 295 (1964)
125 P T Bolwijn, C Th J Alkemade and G A Boschloo, *Phys. Lett.* **4**, 59 (1963)
126 L J Prescott and A Van der Ziel, *Phys. Lett.* **12**, 317 (1964)
127 C Freed and H A Haus, *Appl. Phys. Lett.* **6**, 85 (1965)
128 B Van der Pol, *Phil. Mag.* **3**, 65 (1927)
129 W E Lamb jr, *Phys. Rev.* **134**, A1429 (1964)
130 J A Armstrong and A W Smith, *Phys. Rev. Lett.* **14**, 68 (1965)
131 J A Armstrong and A W Smith, in *Proc. Int. Conf. on the Physics of Quantum Electronics, Puerto Rico, 1965* edited by P Kelley, B Lax and P Tannenwald (McGraw-Hill: New York, 1965) p701
132 A W Smith and J A Armstrong, *Phys. Lett.* **16**, 38 (1965)
133 J A Armstrong and A W Smith, *Phys. Rev.* **140**, A155 (1965)
134 J E Geusic, H M Marcos and L G van Uitert, in *Physics of Quantum Electronics* edited by P L Kelley, B Lax and P E Tannenwald (McGraw-Hill: New York, 1966) p725
135 T P McLean and E R Pike, *Phys. Lett.* **15**, 318 (1965)
136 J Marguin, R Marcy, G Hepner and G Pircher, *Comptes Rendus* **260**, 1361 (1965)
137 F A Johnson, T P McLean and E R Pike, in *Physics of Quantum Electronics* edited by P L Kelly *et al* (McGraw-Hill: New York, 1966) p706
138 C Freed and H A Haus, in *Physics of Quantum Electronics* edited by P L Kelley *et al* (McGraw-Hill: New York) p715
139 H A Haus, *Quarterly Progress Report, Research Laboratory of Electronics, MIT* 15 April 1964, p49
140 *RRE Newsletter and Research Review* no 4, March 1965; there is a brief communication by T P McLean and E R Pike in which, by using the calculation later presented in note 87, they report on an experiment under way

141 E Jakeman, C J Oliver and E R Pike, *J. Phys. A: Gen. Phys.* **1**, 497 (1968)
142 F T Arecchi, *Phys. Rev. Lett.* **15**, 912 (1965)
143 C Freed and H A Haus, *Phys. Rev. Lett.* **15**, 943 (1965)
144 W Martienssen and E Spiller, *Am. J. Phys.* **32**, 919 (1964). Later these authors (*Phys. Rev. Lett.* **16**, 531 (1966)) used a similar system to determine the probability densities for the light intensity of chaotic light if the detector averages over N cells of phase space.
145 F T Arecchi, E Gatti and A Sona, *Phys. Lett.* **20**, 27 (1966)
146 G Farkas, L. Janossy, Z Náray and P. Varga, *Acta Phys. Acad. Sci. Hung.* **18**, 199 (1965); Yu F Skachkov, *JETP* **21**, 1026 (1965)
147 R Jones, F A Johnson, T P McLean and E R Pike, *Phys. Rev. Lett.* **16**, 589 (1966)
148 C Freed and H A Haus, *IEEE J. Quantum Electron.* **QE-2**, 190 (1966)
149 F T Arecchi, A Berne, A Sona and P Burlamacchi, *IEEE J. Quantum Electron.* **QE-2**, 341 (1966)
150 See note 52, W G Wagner and G Birnbaum, *J. Appl. Phys.* **32**, 1185 (1961); J A Fleck jr, *J. Appl. Phys.* **34**, 2997 (1963); R V Pound, *Ann. Phys., NY* **1**, 24 (1957); J Weber, *Rev. Mod. Phys.* **31**, 681 (1959); M P W Strandberg, *Phys. Rev.* **106**, 617 (1957); D E McCumber, *Phys. Rev.* **130**, 675 (1962); W H Wells, *Ann. Phys., NY* **12**, 1 (1961); G Kemeny, *Phys. Rev.* **133**, A69 (1964); F Schwabl and W Thirring, *Erg. Naturw.* **36**, 219 (1964); however, some authors developed non-linear cases, neglecting spontaneous emission, H Haken and H Sauermann, *Z. Phys.* **173**, 261 (1963); **176**, 47 (1963); R M Bevensee, *J. Math. Phys.* **5**, 308 (1964)
151 H Haken, *Z. Phys.* **181**, 96 (1964); **182**, 346 (1965); *Phys. Rev. Lett.* **13**, 329 (1964)
152 See note 129, Lamb had previously circulated his ideas; see for example M Sargent III and M O Scully in *Laser Handbook* edited by F T Arecchi and E O Schultz-DuBois (North-Holland: Amsterdam, 1972) p45
153 J A Armstrong and A W Smith, *Phys. Lett.* **19**, 650 (1965)
154 H Risken, *Z. Phys.* **186**, 85 (1965)
155 A W Smith and J A Armstrong, *Phys. Lett.* **19**, 650 (1966); *Phys. Rev. Lett.* **16**, 1169 (1966)
156 F T Arecchi, G S Rodari and A Sona, *Phys. Lett.* **25A**, 59 (1967); D Meltzer, W Davis and L Mandel, *Appl. Phys. Lett.* **17**, 242 (1970); E Jakeman, C J Oliver, E R Pike, M Lax and M Zwanziger, *J. Phys. A: Gen. Phys.* **3**, L52 (1970)
157 M Scully and W E Lamb jr, *Phys. Rev. Lett.* **16**, 853 (1966); *Phys. Rev.* **159**, 208 (1967); *Phys. Rev.* **166**, 246 (1968)
158 G Lachs, *Phys. Rev.* **138**, B1012 (1965)
159 L Mandel and E Wolf, *Phys. Rev.* **149**, 1033 (1966)
160 H Morawitz, *Phys. Rev.* **139**, A1072 (1965); *Z. Phys.* **195**, 20 (1966)
161 J Peřina, *Phys. Lett.* **24A**, 333 (1967); see also J Peřina, *Coherence of Light* (Van Nostrand: London, 1972); E R Pike, *Riv. Nuovo Cim.* **1**, 277 (1969)
162 For example dead-time effects were treated in note 147 but see also G Bedard *Proc. Phys. Soc.* **90**, 131 (1967). A summary of all the problems

encountered in photon counting measurements can be found in *Photon Correlation and Light Beating Spectroscopy* edited by H Z Cummins and E R Pike, NATO-ASI Series B3 (Plenum: London, 1974) and *Photon Correlation Spectroscopy and Velocimetry* edited by H Z Cummins and E R Pike, NATO-ASI Series B23 (Plenum: London, 1977)

163 Quasi-thermal and laser photon counting distributions compared directly on an oscillograph were published in Professor Kastler's revision of Bruhat's text *Thermodynamique*, Paris (1966); see also E R Pike, *Riv. Nuovo Cim.* **1**, 277 (1969)

164 F A Johnson, R Jones, T P McLean and E R Pike, *Opt. Acta* **14**, 35 (1967)

165 H Gamo, R E Grace and T J Walter, *2nd Rochester Conf. on Coherence and Quantum Optics, Rochester, NY, June 1966*, Conference Abstracts, p183

166 R F Chang, R W Detenbeck, V Korenman, C O Alley jr and U Hochuli, *Phys. Lett.* **25A**, 272 (1967)

167 F Davidson and L Mandel, *Phys. Lett.* **27A**, 579 (1968)

168 F Davidson and L Mandel, *Phys. Lett.* **25A**, 700 (1967)

169 D Meltzer, W Davis and L Mandel, *Appl. Phys. Lett.* **17**, 242 (1970)

170 E Jakeman, C J Oliver, E R Pike, *J. Phys. A: Gen. Phys.* **3**, L45 (1970)

171 See L Brillouin, *Les Statistiques Quantiques* (Les Presses Universitaires France: Paris, 1930) ch 6. For the tendency of a boson gas to form clusters in space see G E Uhlenbeck and L Gropper, *Phys. Rev.* **41**, 79, (1932); F London, *Phys. Rev.* **54**, 947 (1938); *J. Chem. Phys.* **11**, 203 (1943)

172 R H Dicke, *Phys. Rev.* **93**, 99 (1954)

173 B L Morgan and L Mandel, *Phys. Rev. Lett.* **16**, 1012 (1966)

174 Photon bunching is a typical case in which the *P* representation is missing. See for example J Peřina, *Prog. in Opt.* **18**, 129 (1980)

175 H J Kimble, M Dagenais and L Mandel, *Phys. Rev. Lett.* **39**, 691 (1977); *Phys. Rev.* A **18**, 201; M Dagenais and L Mandel, *Phys. Rev. A* **18**, 2217 (1978); see also E Jakeman, E R Pike, P N Pusey and J M Vaughan, *J. Phys. A: Math. Gen.* **10**, L257 (1977); M M Tehrani and L Mandel, *Phys. Rev.* A **17**, 677, 694 (1978); L Mandel, *Opt. Lett.* **4**, 205 (1979). For a general review of these interesting effects see J Peřina, *Prog. Opt.* **18**, 129 (1980)

176 See the general review papers given in notes 121, 161, 162 and 175 and B Saleh, *Photoelectron Statistics* (Springer: Berlin, 1978)

Subject Index

Author Index

The numbers in brackets refer to Notes